POWDER METALLURGY

Edited by **Katsuyoshi Kondoh**

Powder Metallurgy
Edited by Katsuyoshi Kondoh

Published by InTech
Janeza Trdine 9, 51000 Rijeka, Croatia

Publishing Process Manager Daria Nahtigal

Technical Editor Teodora Smiljanic

Cover Designer InTech Design Team

Additional hard copies can be obtained from orders@intechopen.com

Powder Metallurgy, Edited by Katsuyoshi Kondoh
p. cm.
ISBN 978-953-51-0071-3

For EU product safety concerns:
IN TECH d.o.o., Prolaz Marije Krucifikse Kozulić 3, 51000 Rijeka, Croatia,
info@intechopen.com or visit our website at intechopen.com.

Meet the editor

Professor Dr. Katsuyoshi Kondoh is a vice director of Joining and Welding Research Institute (JWRI), Osaka University. He received both B.S./M.S. degrees and PhD from Osaka University majoring in arc welding process. His major research field is powder metallurgy materials and processing, in particular non-ferrous materials and their composites such as magnesium, titanium, aluminum and copper. He also studies on high-performance carbon nanotubes reinforced metal matrix composites and CNTs coating process for mechanical materials and biomaterials. He published more than 200 papers and has about 120 patents.

Contents

Preface

From high-performance, economical and environmental points of view, *Powder metallurgy* process shows remarkable advantages in production of parts and components due to their special compositions by elemental mixing and 3-dimensional near net shape forming methods. Powder metallurgy process can be applied to not only metal materials but also ceramics and organic materials, which both are employed as structural and electrical products. Author contributions to *Powder metallurgy* present excellent and significantly important research topics to evaluate various properties and performance of P/M materials for applying these materials as actual components. In particular, the life estimation of P/M ferrous materials by sliding contact fatigue test and tribological performance evaluation of P/M semi-metallic materials are focused and introduced in this book.

Katsuyoshi Kondoh
Joining and Welding Research Institute (JWRI)
Osaka University
Japan

1

Selection of Best Formulation for Semi-Metallic Brake Friction Materials Development

Talib Ria Jaafar[1], Mohmad Soib Selamat[1] and Ramlan Kasiran[2]
[1]*Advanced Materials Centre, SIRIM Berhad, 34, Jalan Hi-Tech 2/3, Kulim Hi-Tech Park, Kulim,*
[2]*Faculty of Mechanical Engineering, University Technology MARA, ShahAlam*
Malaysia

1. Introduction

Brake friction materials play an important role in braking system. They convert the kinetic energy of a moving car to thermal energy by friction during braking process. The ideal brake friction material should have constant coefficient of friction under various operating conditions such as applied loads, temperature, speeds, mode of braking and in dry or wet conditions so as to maintain the braking characteristics of a vehicle. Besides, it should also posses various desirable properties such as resistance to heat, water and oil, has low wear rate and high thermal stability, exhibits low noise, and does not damage the brake disc. However, it is practically impossible to have all these desired properties. Therefore, some requirements have to be compromised in order to achieve some other requirements. In general, each formulation of friction material has its own unique frictional behaviours and wear-resistance characteristics.

Friction material is a heterogeneous material and is composed of a few elements and each element has its own function such as to improve friction property at low and high temperature, increase strength and rigidity, prolong life, reduce porosity, and reduce noise. Changes in element types or weight percentage of the elements in the formulation may change the physical, mechanical and chemical properties of the brake friction materials to be developed (Lu, 2006; Cho et Al., 2005; Mutlu et al., 2006 & Jang et al., 2004). Earlier researchers have concluded that there is no simple correlation between friction and wear properties of a friction material with the physical and mechanical properties (Tanaka et al., 1973; Todorovic, 1987; Hsu et al. 1997 & Talib et. al, 2006). Therefore, each new formulation developed needs to be subjected to a series of tests to evaluate its friction and wear properties using brake dynamometer as well as on-road braking performance test to ensure that the brake friction material developed will comply with the minimum requirements of its intended application.

Two major types of brake dynamometers are commonly used to evaluate the friction and wear characteristics of the friction materials are the inertia dynamometer and CHASE dynamometer. Inertia dynamometer is used to evaluate a full size brake lining material or brake system by simulating vehicles braking process but it is time consuming and more expensive. On a smaller scale, CHASE dynamometer features low capital expenditure and shorter test time (Tsang, 1985). Chase machine uses a small sample of friction material with a size of 1 inch x 1 inch x 0.25 inch. These brake dynamometers has been used to tests friction materials for quality control, lining development and friction materials property assessments in a lab scale rather then having a series of vehicle tests on a test track or road (Sander, 2001).

The two main types of tests used to evaluate the performance under different loading, speed, temperature and pedal force are, namely, inertia-dynamometer and vehicle-level testing. Inertia-dynamometer test procedures or vehicle testing simulation is used as a cost-effective method to evaluate brake performance in a laboratory-controlled environment. The automotive industry uses inertia-dynamometer testing for screening, development and regular audit testing. Blau postulated that there is no laboratory wear test of vehicle brake materials can simulate all aspects of a brake's operating environment (Blau, 2001). Vehicle testing on the test track is the ultimate judge for overall brake performance testing and evaluation.

Generally, in normal life we cannot avoid friction phenomenon. It still happens as long as there is a relative motion between two components. Even though friction can cause wear of materials, sometimes the process of friction is required such as in the brake system, clutch, and grinding. During a braking process, brake pads or brake shoes are pressed against the rotating brake disc or drum. During this process the friction materials and the brake disc are subjected to wear.

Friction is a continuous process but wear is a more complicated process than friction because it involves plastic deformation plus localised fracture event (Rigney, 1997), microstructural changes (Talib et al., 2003), and chemical changes (Jacko, 1977). Wear process in dry sliding contacts begins with particle detachment from the contact material surface due to formation of plastic deformation, material transfers to the opposite mating surface and formation of mechanical alloyed layers (Chen & Rigney 1985), finally elimination of wear fragments from the tribosystem as the wear debris. Wear mechanism in the operation during braking is a complex mechanism and no single mechanism was found to be fully operating (Rhee, 1973 & 1976; Bros & Sciesczka, 1977; Jacko et al., 1984; Talib et. al. 2007) and the major wear phenomena observed during braking processes were; (i) abrasive (ii) adhesive (iii) fatigue (iv) delamination and (v) thermal wear.

Friction and wear characteristics of friction material play an important role in deciding which new formulations developed are suitable for the brake system. The friction and wear behaviours of automotive brake pads are very complex to predict which depend on the various parameters such as microchemical structure of the pad and the metallic counter-face, rotating speed, pressure and contact surface temperature (Ingo et al., 2004). Composition and formulation of brake pads also play a big role on the friction behaviour, and since composition-property relationship are not known well enough, the formulation task is based on trial and error and thus is expensive and time consuming (Österl & Urban, 2004). Generally brake pads have a friction coefficient, μ between 0.3 and 0.6 (Blau, 2001).

In this work, ten (10) new friction material formulations which are composed of between eight (8) to foureen (14) elements have been developed using power metallurgy technique. In addition, a commercially-available brake pad, labelled as COM, was chosen for comparison purposed. Each sample was subjected to density, hardness, porosity, friction and wear, brake effectiveness and on-road braking performance tests in accordance with various relevant international standards. The best formulation was selected based on the following methodology;

i. First screenings - screening of the developed formulations based on the results of the physical and mechanical tests.
ii. Second screening – screening of the developed formulations based on the results of friction and wear tests performed on CHASE brake lining friction machine.
iii. Third screening - screening of the developed formulations based on the results of brake dynamometer tests.
iv. Final selection - selection of the best developed formulations is based on the compliance with the on-road braking performance requirements.

Correlation among the mechanical, tribological and performance will also be discussed in this work. Wear phenomena on the worn surface after on-the road performance test will be examined and postulated.

2. Materials and method

2.1 Semi-metallic friction materials

Ten semi-metallic brake pad formulations which composed of between eight (8) to fourteen (14) ingredients were produced in this study using powder metallurgy route (Table 1). The powder metallurgy route consists of the following processes, namely, (ii) dry mixing, (ii) preparation of backing plate, (iii) pre-form compaction (iv) hot compaction, (v) post-baking, and (vi) finishing. The prototype samples were marked as SM1, SM2, SM3, SM4, SM5, SM6, SM7, SM8, SM9 and SM10. Figure 1 shows two (2) example microstructure of the newly developed semi-metallic brake pad. It can be seen that the brake pads developed are not a homogenous material. The particle size of each element is not unifrom in size and the distribution of the element is not well dispersed in the matrix.

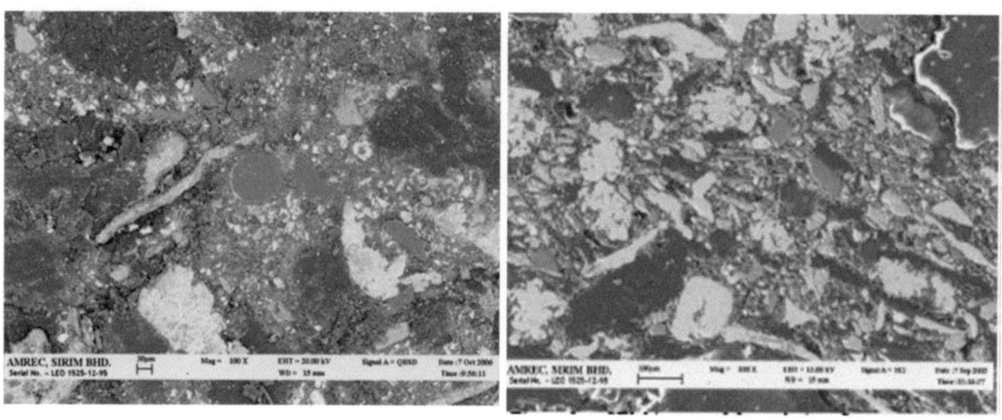

(a) (b)

Fig. 1. Surface morphology of semi-metallic brake pad; (a) sample SM1, (b) sample SM4

Ingredients	Formulation (Weight %)									
	SM1	**SM2**	**SM3**	**SM4**	**SM5**	**SM6**	**SM7**	**SM8**	**SM9**	**SM10**
Resin	10.0	10.0	9.0	9.0	9.0	9.0	9.0	8.0	12.0	9.0
Kevlar	-	-	2.0	-	2.0	2.0	-	-	-	3.0
Steel fiber	20.0	23.0	31.0	20.0	30.0	31.0	22.0	24.0	22.0	25.0
Organic fiber	5.0	-	-	10.0	2.0	2.0	10.0	8.0	7.0	5.0
Copper fiber	-	-	2.0	6.0	2.0	3.0	6.0	-	8.0	3.0
Graphite	16.0	19.0	7.0	13.0	15.0	7.0	13.0	11.0	16.0	6.0
Antimony	-	-	3.0	-	3.0	3.0	-	-	5.0	
Iron oxide	34.0	24.0	18.0	9.0	16.0	18.0	9.0	15.0	3.0	21.0
Novacite silica	-	3.0	-	3.0	-	-	3.0	2.0	6.0	3.0
Alumina oxide	-	2.0	1.0	5.0	2.0	1.0	2.0	2.0		2.0
Zinc oxide	1.0	-	1.0	-	2.0	-	-	3.0	-	2.0
Rubber	-	3.0	4.0	3.0	3.0	3.0	3.0	3.0	3.0	5.0
White rock	-	-	2.0	3.0	2.0	3.0	3.0	6.0	3.0	3.0
Barium	8.0	10.0	20.0	19.0	8.0	18.0	20.0	14.0	-	7.0
Friction dust	6.0	6.0	-	-	4.0	-	-	4.0	15.0	6.0
TOTAL	100	100	100	100	100	100	100	100	100	100

Table 1. Ingredients of semi-metallic brake pad

2.2 Physical and mechanical tests

Each sample produced is subjected to density, porosity and hardness tests. Density of semi-metallic brake pads was obtained using Archimedes' principle in accordance with Malaysian Standard MS 474: Part 1: 2003 test procedures. Hardness was measured using a Rockwell hardness tester model Mitutoyo Ark 600 in S scale in accordance with Japanese Industrial Standard JIS 4421: 1996 test procedures. The hardness of the samples is the arithmetic mean of ten measurements. Porosity was obtained in accordance with JIS 4418: 1996 test procedures using a hot bath model Tech-Lab Digital Heating.

2.3 Friction and wear tests

Friction coefficient and wear were results were obtained which is in compliance with Society of Automotive Engineer SAE J661 test procedures. In this test, the sample was pressed against a rotating brake drum with a constant rotating speed of 417 rpm under the load of 647 N and subjected to test program as shows in Table 2. Briefly, each sample was subjected to seven test runs with the following sequences; (i) baseline, (ii) first fade, (iii) first recovery, (iv) wear, (v) second fade, (vi) second recovery, and (viii) baseline rerun. The samples thickness were measured and weighed before and after testing. Friction coefficient and wear tests were conducted by Greening Testing Laboratories Inc., USA using CHASE machine.

Test Sequence	Load (N)	Speed (rpm)	Braking mode
Conditioning	440	312	Continuous braking for 20 mins
Initial thickness & mass measurement	222	208	Continuous braking for 5 mins
Baseline run	667	417	Intermittent braking 10 s ON, 20 s OFF for 20 applications
First fade run	667 .	417	Continuous 10 minutes or until 288 °C is attained which ever come first
First recovery run	667	417	10 seconds application at 260, 204 , 149 and 93 °C
Wear run	667	417	Intermittent 20 s ON, 10 s OFF for 100 applications.
Second fade run	667	417	Continuous 10 mins or until 343 °C is attained which ever come first
Second recovery run	667	417	10 seconds application at 316, 260, 204 , 149, 93 °C
Baseline rerun	667	417	Intermittent 10 s ON, 20 s OFF for 20 applications
Final thickness and mass measurement			Repeat initial thickness and mass measurement

Table 2. Friction and Wear Assessment Test Program

2.4 Brake effective tests

The braking performance of the developed semi-material brake pads were determined using brake dynamometer test in accordance with Society of Automotive Engineers standard SAE J2552 issued in August 1999 (available from SAE, 400 Commonwealth Dr, Warrendale, PA 15096, USA). This standard assesses the effectiveness behavior of a friction material with regard to pressure, temperature and speed. Vehicle brake simulations are conducted on an inertia dynamometer, which simulate kinetic energy of the vehicle mass moving at speed. Before the beginning of performance measurement, a burnishing period for conditioning the lining/counterface pairs require more than 200 conditioning stops. After conditioning, dynamometer-based lining tests were subjected to pressure-sensitive stops, speed-sensitive drags, fade and recovery tests. Table 3, briefly shows the test sequences. Dynamometer global brake effectiveness tests were conducted by Greening Testing Laboratories Inc., USA using single end brake brake dynamometer. Each sample was conducted on a new brake rotor. In this study, the focus is only on the friction coefficient and wear characteristics of the sample.

The purpose of this investigation was to evaluate the performance of the semi-metallic brake pads for Proton WIRA using brake inertia dynamometer. Figure 2 shows prototype brake pad and Proton Wira's drive shaft assembly. The technical specification of the brake effectiveness test is shown in Table 4.

Bil		Snub	Cycle	Speed (km/h)	Pressure (kPa)	Initial temp (^{o}C)
1.	Green µ	30	1	80 to 30	3000	< 100
2.	Burnish	32	6	80 to 30	Varying pressure	< 100
3.	Characteristic 1	6	1	80 to 30	3000	< 100
4.	Speed/press sensitivity	8 8 8 8 8	1 1 1 1 1	40 to 5 80 to 40 120 to 80 160 to 130 200 to 170	Increasing pressure 1000 to 8000	< 100
5.	Characteristic 2	6	1	80 to 30	3000	< 100
6.	Cold	1	1	40 to 5	3000	< 40
7.	Motorway application	1 1	1 1	100 to 5 0.9 V_{max} to 0.5 V_{max}	0.6 g	< 50
8.	Characteristic 3	18	1	80 to 30	3000	< 100
9.	Fade 1	15	1	100 to 5	16000 0.4 g	< 100 < 550
10.	Recovery 1	18	1	80 to 30	3000	< 100
11.	Temp/press Sensitivity 100/80 ^{o}C	8	1	80 to 30	Increasing pressure 1000 to 8000	< 100
12.	Temp/press Ssnsitivity 500/300 ^{o}C	9	1	80 to 30	3000	< 100
	Pressure line 500/300 ^{o}C	8	1	80 to 30	Increasing pressure 1000 to 8000	< 550
13.	Recovery 2	18	1	80 to 30	3000	< 100
14.	Fade 2	15	1	100 to 5	16000 0.4 g	< 100 < 550
15.	Recovery 3	18	1	80 to 30	3000	< 100

Table 3. Brake dynamometer test sequence

Item	Specification
Vehicle System Simulated	1996 Proton Wira 1.5 GL Front
Brake Configuration	single piston, separate function disc brake
Piston Diameter	54 mm
Rotor Diameter x Thickness	236 x 18 mm
Rotor Mass (nominal)	3.7 kg
Rotor Effective Radius	95.88 mm
Axle Load	830 kg
Test Inertia	34.7 kg m^2
Static Loaded Radius / Rolling Radius	287.02 mm
Simulated Wheel Load	421 kg
Wheel Rotation	: right hand

Table 4. Technical specifications of the brake effective test

Fig. 2. Semi-metallic brake pad and front axle brake system

2.5 On-road performance tests

In the road performance test, the brake pads were fitted to the brake system of PROTON WIRA 1.5GL with the following test conditions: (i) unladen vehicle, (ii) disconnected engine, (iii) tire inflated to the manufacturer's specifications, (iv) the road was hard, level and dry, (vi) the wind speed was below 5 m/s. The road performance test divided into three types, namely; (i) cold effectiveness test, (ii) heat fade test, and (iii) recovery test.

The on-road braking performance tests on car were performed following closely procedure described in the ECE R13, Annex 3. Modification on the procedure was necessary due to the limitation of the test track conditions (Table 5). Real application testing of the friction materials were carried out using Proton Wira 1.5 (Table 6). Figure 3 shows the test equipment set-up.

PARAMETER	ECE R13	MODIFIED ECE R13
<u>Type-O: Cold brake</u>		
• Initial vehicle speed	120 km/h	100 km/h
• Brake pedal force	65 – 500 N	65 – 500 N
• Engine disconnected	Yes	Yes
• Average temperature	65 – 100 °C	65 – 100 °C
• Vehicle must be laden & unladen	Yes	2 people
<u>Type-1: Fade test</u>		
Heating procedure:		
• Braking speed	120 - 60 km/h	100 - 50 km/h
• Brake pedal force	Equivalent to 3 m/s^2 deceleration	50% of Type-O
• No. of brake application	15	12
• Vehicle must be laden	Yes	2 people
• Engine connected	Yes	Yes
Hot performance:		
• Initial vehicle speed	120 km/h	100 km/h
• Brake pedal force	Same force obtained in Type-O test	Same force obtained in Type-O test
• Engine disconnected	Yes	Yes
Recovery procedure:		
• Braking speed	50 - 0 km/h	50 - 0 km/h
• Brake pedal force	Equivalent to 3 m/s^2 deceleration	50% of Type-O
• No. of stops with 1.5 km interval	4	4
• Vehicle must be laden	Yes	2 people
• Engine connected	Yes	Yes
Recovery performance:		
• Initial vehicle speed	120 km/h	100 km/h
• Brake pedal force	Same force obtained in Type-O test	Same force obtained in Type-O test
• Engine disconnected	Yes	Yes

Table 5. Modified ECE R13 test procedure

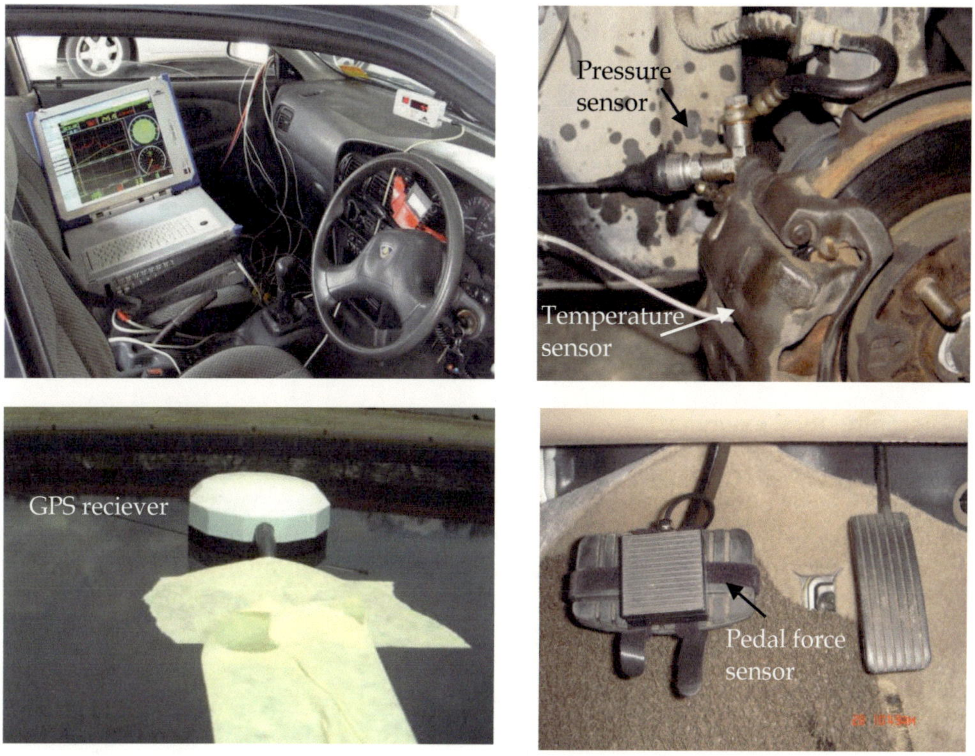

Fig. 3. Test equipment set-up; (a) *Dewetron DEWE-5000* system, (b) pressure sensor and thermocouple, (c) GPS receiver installed on the roof (d) pedal force sensor.

Manufcaturer	PROTON
Model	Proton WIRA 1.5S
Engine capacity	1,468 cc
Max. Power	66 kW
Max Torque	126NM @ 3000 rpm
Gear system	manual
Wheel size	175/70/R 13
Tire pressure	190 kPa

Table 6. Test car specifications

In the cold test, the test was conducted with the brake lining temperature below 100 °C prior to each brake application and comprised of six brakings including familiarization. The brake test was carried out at the initial vehicle speed of 100 km/hr. The test data such as vehicle speed, lining temperature and braking distance were recorded using a brake measuring system from *Dewetron* model DEWE5000. If the pedal force applied is more than 500 N for wheel locking to occur, the brake pad is considered fail to comply with the requirements and the next test (heat fade and recovery test) will not be conducted.

Fade test is used to evaluate the brake performance under high brake material temperature. Prior to this test, the service brake of the test car was heated by successively applying the brake. The initial speed at beginning of this heating procedure was set at 100 km/hr and speed at the end of braking was set at 50 km/hr with brake pedal force capable of generating 50 % type-O deceleration. This process was repeated for 12 brake applications. Upon completing this heating procedure, the test vehicle was accelerate to initial vehicle speed of 100km and brake was applied using the same pedal force as in cold effective test of that particular sample. Immediately, the recovery test was conducted with the following test procedure; (a) make four stops from 50 km/hr with the same pedal force applied during heating process of heat fade test. Immediately after each stop, accelerate the vehicle to 50 km/hr and make subsequent stop, (b) accelerate the test vehicle to a speed of 100km and then brake pedal was applied with the same pedal force cold test.

2.6 Microstructural examination

The worn surface after on-road performance test were analyzed using scanning electron microscope model Leo equipped with a Oxford energy-dispersive X-ray analyzer (EDX).The samples for microstructural examination were cut from a real-size brake pad and coat with platinum using sputter coater to ovoid charging effect during the analysis.

3. Results and discussion

3.1 Physical and mechanical properties

Specific gravity is the relative density of a substance compared to the density of pure water and porosity is the percentage of pore volume with the bulk total volume. Hardness is a measure of material resistance to plastic deformation. The specific gravity, porosity and hardness properties depend on the ingredients and weight percentage used as well as the manufacturing process parameters. Test results of physical and mechanical properties are shown in Table 7 and Figure 4. Ideally, the highest specific gravity should give the lowest porosity and highest hardness reading. But in friction materials this postulation does not apply as shown in this investigation. For example, sample SM6 has the highest specific gravity reading but does producing the highest hardness result. Highest hardness reading was recorded by sample SM1, but this sample is not producing the highest density and the lowest porosity reading (Table 7).

Figure 4 exhibits the correlation among the specific gravity, porosity and hardness. Figure 4a indicates that the specimens with high porosity tend to exhibit low specific gravity. However, Figure 4b shows that there is no simple correlation between hardness with porosity. Brake pad should have a certain amount of porosity to minimize the effect of water and oil on the friction coefficient and to reduce the brake noise. Hardness should be decreased as much as is feasible to increase performance stability and the steel fiber content should be less than 7% in order to reduce rotor thickness variation (Sasaki, 1995). He also found that increasing porosity by more than 10 % could reduce the brake noise. But if the porosity too much, the hardness will be reduced resulting increase in wear rate of friction material. Friction material is not a homogeneous material, when the indenter hits on the metallic component the hardness will be higher, otherwise when it hits on polymeric component the hardness will be lower (Talib et al., 2008). Thus, the hardness of the friction material is not a representative of the bulk property.

Ideally the friction materials developed should have the best physical and mechanical properties in order to get the best brake effective performance. In case of friction materials, this phenomenon does not apply (Todorovic, 1987; Filip et al., 1995; Talib et al., 2006). The physical and mechanical properties of friction material can not be predicted based on type of ingredient used, particle size and shape, weight percentage of the ingredient. It also depends on manufacturing process parameters such as powder mixing duration, compaction pressure, compaction duration, degassing time, and post curing temperature and time. Based on the above observations, the best formulation can not be selected using physical and mechanical properties. The physical and mechanical properties could be used to control the quality of the formulations that has been developed during the manufacturing process. Consistent physical and mechanical properties of the same formulation reveal that the friction material manufacturing process is in control.

Bil	Sample	Specific gravity	Porosity (%)	Hardness (HRS)
1.	COM	2.69	10.4	69.9
2.	SM1	2.76	7.9	85.1
3.	SM2	2.73	2.6	83.8
4.	SM3	2.36	24.2	76.7
5.	SM4	2.30	21.4	70.6
6.	SM5	2.27	20.2	50.0
7.	SM6	3.25	2.5	69.7
8.	SM7	2.75	7.7	69.6
9.	SM8	2.65	9.5	60.5
10.	SM9	2.32	3.1	73.0
11.	SM10	2.43	16.1	51.1

Table 7. Physical and mechanical test results

3.2 Friction and wear properties

Friction material is a heterogeneous material and composed of a few elements. Therefore, the selection of material and weight percentage used in the friction formulation will significantly affect the tribological behaviour of the brake pad [Hoyer et al.1999]. Society of Automotive Engineer introducing two letter codes in classifying the friction material, where first letter represents normal friction coefficient and the second letter represents hot friction coefficient [SAE J886] as shown in Table 8. Normal friction coefficient is defined as average of the four readings taken at 200, 250, 300 and 400°F on the second fade curve. The hot friction coefficient is defined as the average of the ten readings taken at 400 and 300°F on the first recovery; 450, 500, 550, 600 and 650°F of the second fade; and 500, 400 and 300°F of the second recovery run. Figure 5 shows sample SM5 CHASE test results.

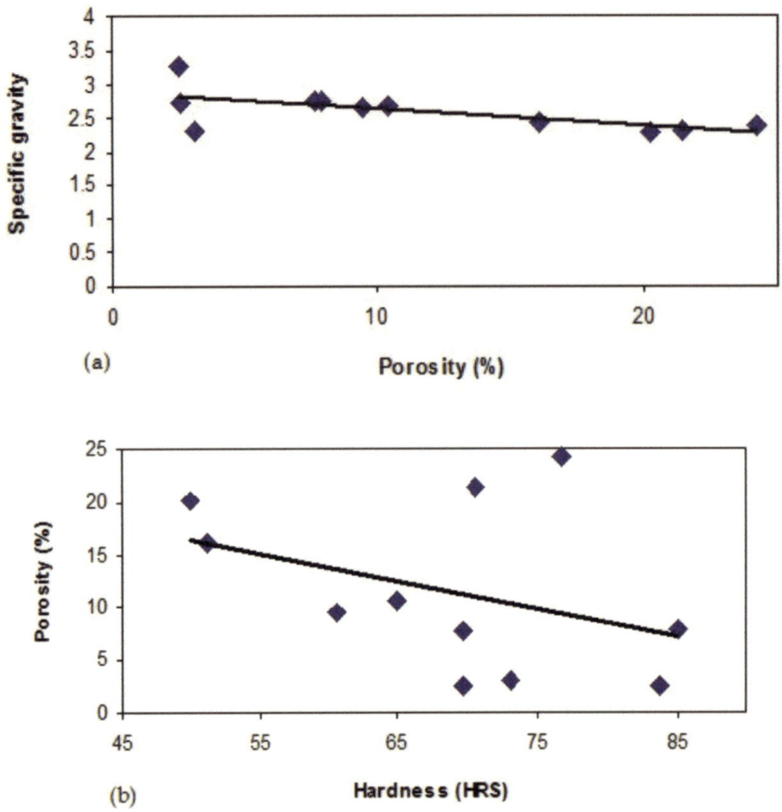

Fig. 4. The relationship between physical properties of friction material used in this study: (a) specific gravity Vs. porosity, (b) hardness Vs. porosity.

Class code	Coefficient of friction
C	Below 0.15
D	Over 0.15 - not over 0.25
E	Over 0.25 - not over 0.35
F	Over 0.35 - not over 0.45
G	Over 0.45 - not over 0.55
H	Over 0.55
Z	unclassified

Table 8. SAE Recommended Practice J866 list for codes and associated friction coefficient

Friction and wear characteristics of friction material play an important role in deciding which new formulations developed are suitable for the brake system designed for a particular vehicle. CHASE test is used in a laboratory for screening of new material formulations prior to inertia dynamometer tests based on friction and wear test results. In deciding which samples are to be subjected to dynamometer tests, the following requirements were set; (a) shall have normal friction coefficient of class E and above, or a hot of class D and above, (b) shall have friction coefficient above 0.15 between 200 and 550 ^{O}F

inclusive in second fade, or between 300 and 200 °F during the secondary fade. These requirements are in line the requirement set by Automotive Manufacturer Equipment Companies Agency, USA.

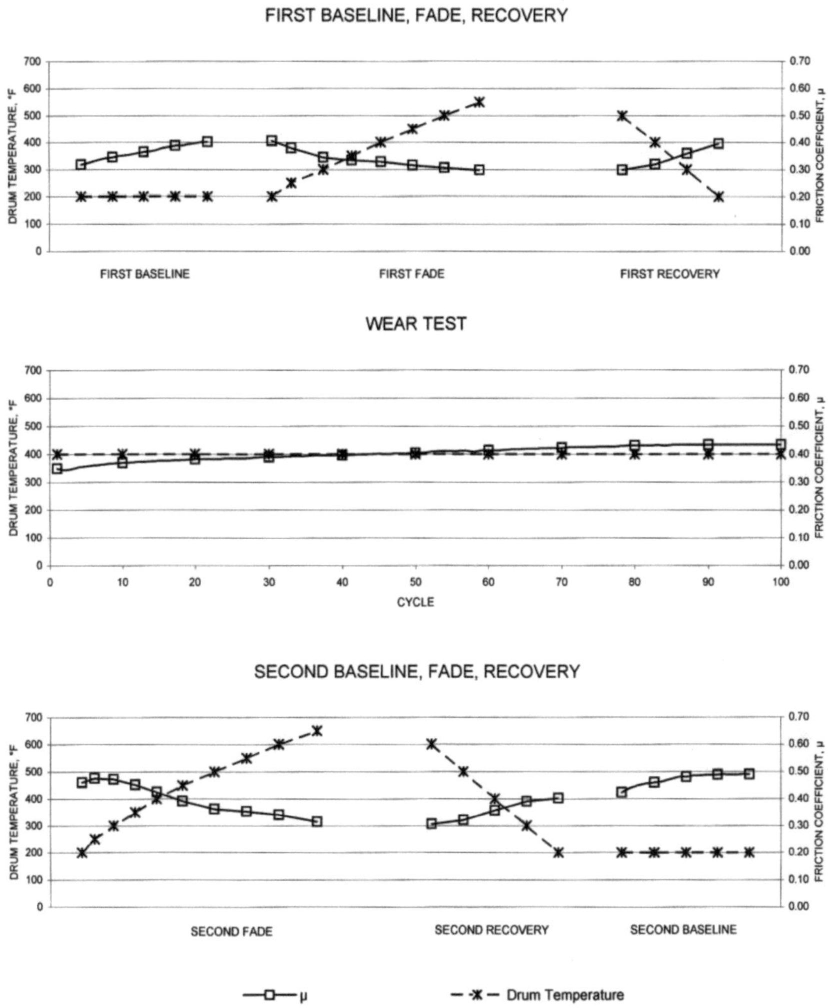

Fig. 5. Friction coefficient characteristic of sample SM5

Test results of friction and wear assessment tests are shown in Table 9 and Figure 6. Analysis of test results showed that all samples developed met with the requirements. During braking, the accumulation of heat will cause high surface temperature on the brake lining materials. The degradation of the polymer materials may cause brake fade in which the friction is reduced as the temperature increased [Begelinger et al. 1973; Rhee 1971]. The ensuing reduction of friction coefficient may also be explained by the shearing of the peak asperities and formation of friction during braking process [Talib et al. 2007]. All hot friction coefficients of the prototype samples reduced except sample SM8. This phenomenon may be due to the formation of metallic layer [Talib, 2001; Scieszka, 1980; Lim et al., 1987] and carbon layer [Begelinger et al., 1973; Zhigao & Xiaofie, 1991]. Sample SM1 and SM2 have lower average thickness loss than the commercial sample. The other eight samples have higher average thickness loss. Even though, higher wear loss resulted in shorter life, the samples which higher average thickness loss will also be subjected to brake effective dynamometer so that correlation thickness loss between these two tests could be made.

Sample	Normal Friction		Hot Friction		Thickness loss (mm)
	μ	Code	μ	Code	
COM	0.422	F	0.385	F	2.79
SM1	0.385	F	0.316	E	0.51
SM2	0.450	F	0.383	F	0.76
SM3	0.459	G	0.352	F	1.52
SM4	0.471	G	0.373	F	2.79
SM5	0.457	G	0.360	F	1.02
SM6	0.417	F	0.345	E	2.03
SM7	0.438	F	0.430	F	3.31
SM8	0.532	G	0.544	G	6.09
SM9	0.374	F	0.322	E	1.27
SM10	0.554	H	0.458	G	5.84

Table 9. Friction and Wear Assessment Test Results

Figure 7 shows the relationship between the hardness with the friction coefficient and average wear. Figure 7a indicates that the sample with high hardness tend to exhibit low friction coefficient. More & Tagert (1952) and Mokhtar (1982) likewise concluded that the coefficient of friction decreased with increase in hardness. Generally, the harder samples were supposed to have a lower average thickness loss. But in this investigation, it was found that this postulation does apply with the friction materials (Figure 7b). Thus, it could be concluded that there is no direct correlation between hardness with average thickness wear loss. Filip et al. (1995) reported that hardness of brake lining materials cannot be simply related to the content of structural constituents, and there is no correlation between hardness and wear resistance.

Figure 8 shows the relationship between the friction coefficient, average thickness loss and porosity. Brake pad should have a certain amount of porosity to minimize the effect of water and oil on the friction coefficient. Sasaki (1995) found that increasing porosity by more than 10 % could reduce the brake noise. It was observed that the sample with high porosity tend to exhibit high friction coefficient. On the other hand, Figure 8b, indicates that there is no direct correlation between average thickness loss with porosity.

During braking, the friction materials wear-off due to friction resistance between the friction materials with the counter face material made of grey cast iron. Wear rate of friction depend many factors such as operating parameters (temperature, speed, braking time), mode of braking (continuous, intermittent braking), wear mechanism in operation during braking (adhesion, abrasion, fatigue). When above the degradation temperature (230 ^{o}C), the

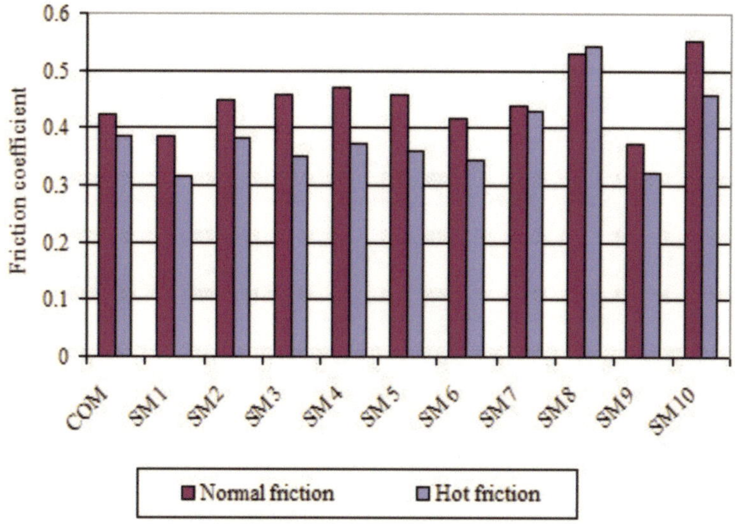

Fig. 6. Normal and hot friction coefficient

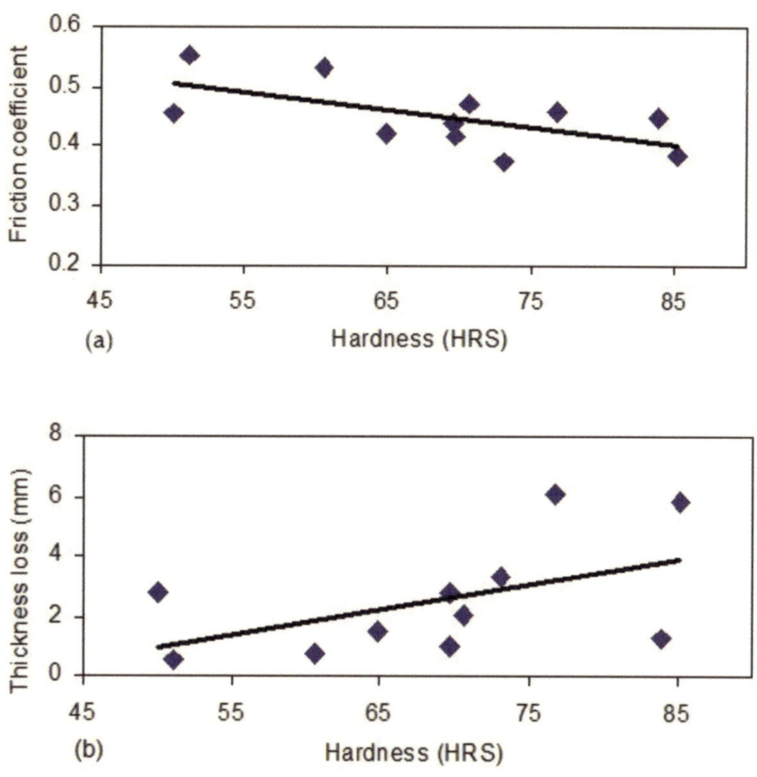

Fig. 7. The relationship between physical properties of friction material used in this study: (a) friction coefficient vs. hardness, (b) average thickness loss Vs. hardness.

binding properties of resin will become weak. As surface temperature increase with increased braking times, the yield strength of the materials will be decreased and leads to change in the wear mechanism and the real contact configuration as well as destruction of friction film. Thus, the wear rate cannot be predicted based on physical and mechanical properties because wear do not depend on material property but rather depends on tribosystem property.

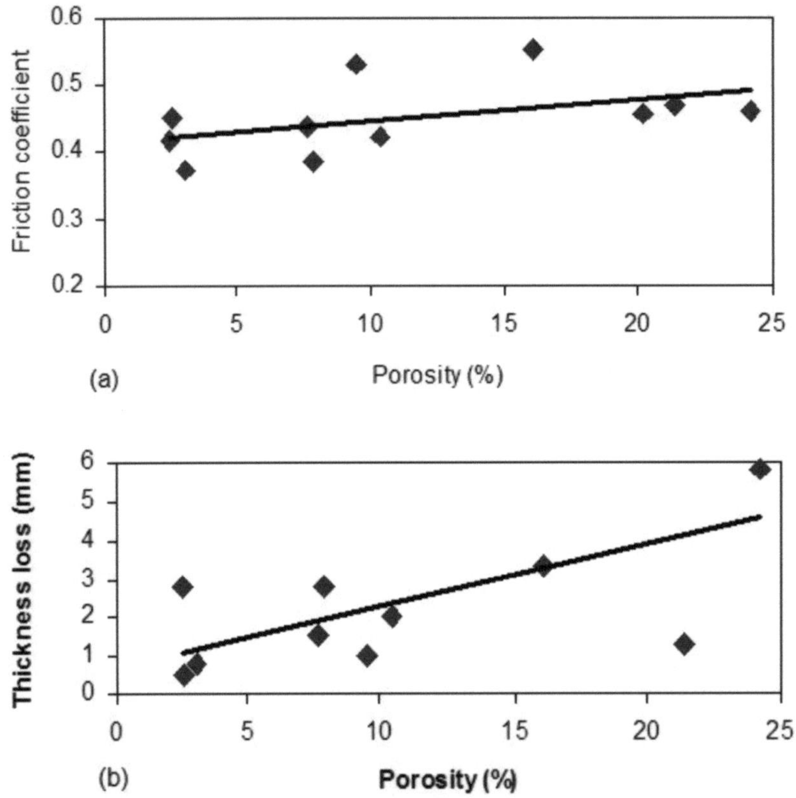

Fig. 8. The relationship between mechanical properties of friction material used in this study: (a) friction coefficient vs. porosity, (b) average thickness loss Vs. porosity.

The friction and wear properties of friction materials depend on a number of different factors such as pressure, speed, interface temperature, composition of friction material and the metal member of the friction pair, duration and length of the friction path, friction material density, its modulus of elasticity, type, design and geometry of friction mechanism [Torovic, 1987]. From the analyses on test data, the following postulation could be made; (i) higher hardness tend to reduce friction coefficient, (ii) higher porosity tend to increase friction coefficient, (iii) there is no simple correlation between average thickness loss with hardness and porosity. So in deciding with formulation can be used in the prototype production, friction and wear test results are the main factor to be considered. Based on test results, it could be concluded that all prototype samples complied with the requirements and will be subjected to brake effective dynamometer tests.

3.2 Brake effective parameters: Friction coefficient and wear

Table 10 shows the friction and coefficient and wear results. Test results show that only sample SM2 and SM9 have a minimum friction coefficient value below than 0.15 during the first fade test sequence. Lower friction coefficient requires a longer braking distance before the vehicle can be stop, which can cause road accident. In this test segment, the temperature was increased from 100 to 550°C under the line pressure of 16 MPa. Under this condition, the brake fade will take place. This fading effect is associated with the decomposition of the organic binder which takes place between 250 and 475 ^{o}C (Ramoussse et al., 2001). The friction coefficient of the friction materials will vary with temperature and will fall off dramatically as the contact temperature exceed the maximum organic decomposition temperature depending on the ingredient and weight percentage used. However, in the second fade test, the friction coefficient of the sample SM2 and SM9 have show a better result which is above 0.15, the minimum requirement. Characteristic friction after second fade also shows almost recover to the characteristic friction in the early stage. Thus all the formulation developed will be subjected to on-road performance test.

During braking process, brake pad is pressed against the brake rotor resulting in wear-off the brake pad as well as the rotor material. Brake pad is designed as the sacrificial element due to it low cost and ease of maintenance. Average values of wear detected after completion of the brake effective dynamometer test procedure is given in Table 10. Wear data are different for different formulation due to different ingredient and weight percentage used in the composition. Wear characteristics is difficult to predict because it depend on the physical, mechanical, chemical characteristic as well as the microstructure changes during the braking process.

Sample	Average friction coefficient				
	Characteristic	First fade	Second fade	Characteristic	Thickness loss (mm)
COM	0.50	0.28	0.29	0.32	1.87
SM1	0.42	0.25	0.25	0.31	1.57
SM2	0.42	0.12	0.34	0.39	1.44
SM3	0.42	0.33	0.28	0.33	1.96
SM4	0.44	0.26	0.29	0.36	1.25
SM5	0.45	0.29	0.29	0.34	1.65
SM6	0.42	0.28	0.31	0.34	3.68
SM7	0.43	0.23	0.28	0.33	2.27
SM8	0.47	0.28	0.35	0.36	4.08
SM9	0.36	0.09	0.18	0.32	1.89
SM10	0.48	0.28	0.32	0.26	3.15

Table 10. Brake dynamometer test results

Figure 9 shows example of friction coefficient characteristics under different stops. The friction coefficient characteristics for other samples vary for different stops as apparent from Figure 9. The first and the second characteristic, first and second fade, and recovery sequences reflect on the performance of brake lining. It can be seen from the fade test, the coefficient of friction decreases with increased in temperature. This is attributed to physical and mechanical, chemical and microstructural changes on the contact surface [Scieszka, 1980; Jacko, 1977; Talib et al., 2003; Ingo et al., 2004].

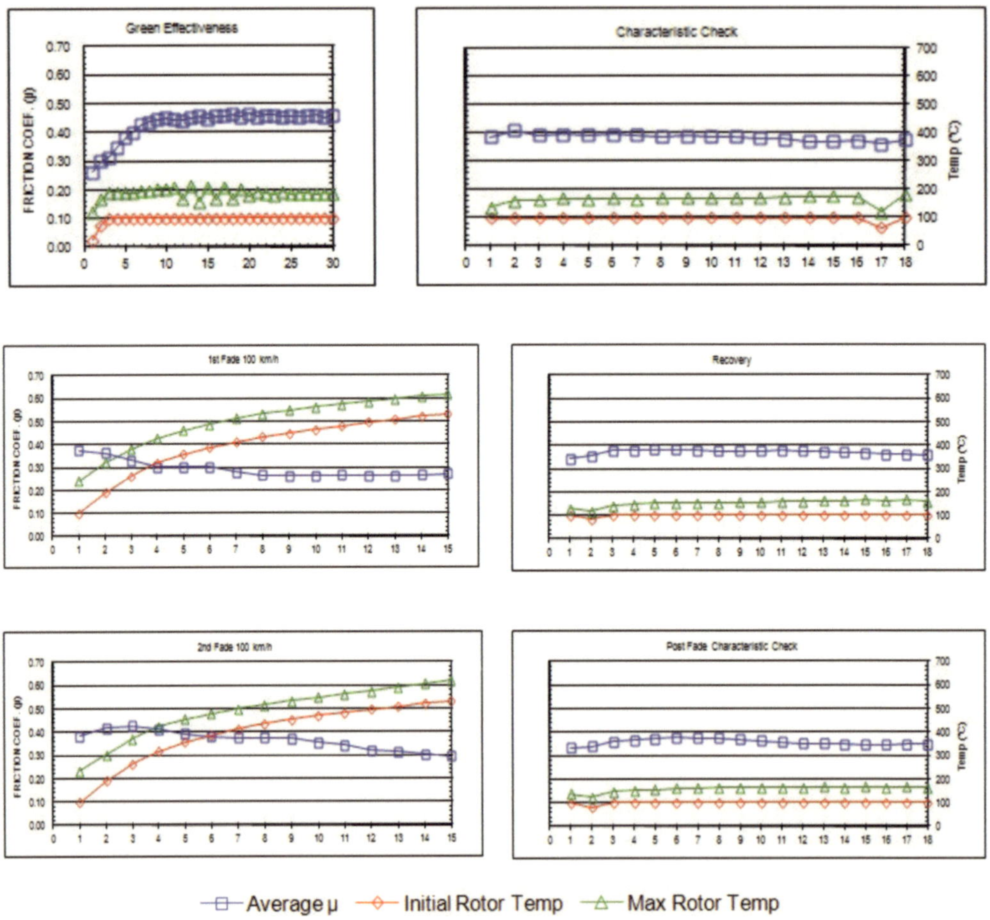

Fig. 9. Characteristic results of brake dynamometer tests

Even though the operating pressure, time and braking sequence of CHASE and dynamometer is not the same, it supposed to produce the friction coefficient results of the same trend for different composition. However, this postulation does not materialise in case of friction materials. It can be seen from Figure 10 that only sample COM, SM1 and SM3 have higher friction coefficient when the samples were subjected to dynamometer tests as compared to CHASE friction coefficient and the variation between friction coefficient reading of CHASE and dynamometer is also not same. Thus, it could be concluded that there is no direct correlation between the friction coefficient between CHASE and dynamometer tests. This was taught due to dependent of friction coefficient with material composition, microstructure and tribosytem.

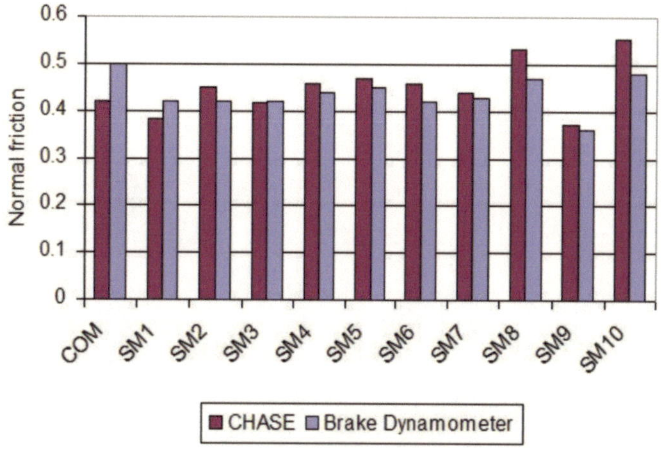

Fig. 10. Bar chart of normal and characteristic friction of CHASE and dynamometer tests.

In case of hot friction, the variation between CHASE and dynamometer results is quite high (Figure 11). This could due to the severity of test conditions applied during dynamometer test. Surface temperature increases when the operating variables such as load, speed and braking time are increased. In the dynamometer test, the surface temperature is increased up to 550°C which much higher compared to CHASE test which is about 300°C. As the surface temperature increases, the polymer materials will degrade. The onset of degradation of the friction material starts at 230 °C, and the degree of degradation increases with temperature within the range of 269 - 400 °C [Zhigao & Xiaofei; 1991]. The degradation of the polymer materials may cause brake fade in which the friction is reduced as the temperature increased [Rhee, 1971; Talib 2001]. The high temperature will also decrease the yield strength and leads to changes in the wear mechanism and the real contact configuration [So, 1996]. These phenomena could be the reason why the friction coefficients during dynamometer test much lower than one during CHASE test. The different between the friction coefficients for particular composition is not the same. This could be due to the heterogeneous properties friction materials. Thus, it could be concluded that there no simple correlation between the friction material under high temperature test condition when subjected to CHASE and dynamometer tests.

Figure 12 shows the data of material thickness losses during braking tests on CHASE machine and brake dynamometer. The results show there is no correlation between the two test results. These variations can be due to the fact that CHASE machine uses a small material sample (i.e. 1 inch x 1 inch x 0.25 inch) pressed against a large rotating drum that does not represent the actual size of the lining material in its real intended application. Whereas, the brake dynamometer evaluates a full size brake lining material as it is in real application and thus simulating the actual braking condition of a vehicle. Thus, CHASE machine is not recommended for evaluation of the thickness loss of the developed sample in full size application. Ideally, all friction materials shall be tested and evaluated in all conditions that they may encounter during their service such as under various brake operating parameters (load, temperature and braking duration), road conditions (downhill and winding roads) and wheather conditions (rain, sunshine and snow). For all these, different vehicles will require different friction materials and unfortuantely, CHASE machine perform rather poorly in predicting the actual performance of the materials when they are put into real life application.

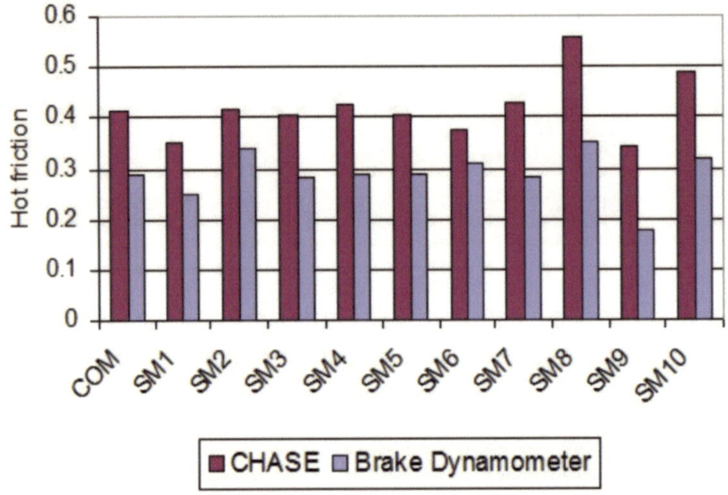

Fig. 11. Bar chart of thickness loss during CHASE and dynamometer test.

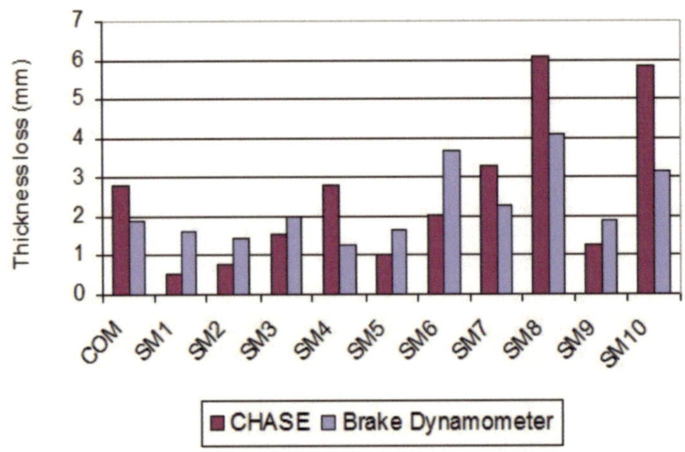

Fig. 12. Bar chart of thickness loss during CHASE and dynamometer test.

3.3 On-road performance

The development and validation of a friction material involve a significant amount of testing in laboratory and on the road. As a vehicle is typically used under various road and driving conditions, a friction material shall be tested in conditions closely representing these driving conditions. Brake friction material developers will look for quantitative data from these tests to evaluate their material formulations and track the effects of the modifications that are made during the course of the product development. Of all the tests carried out during a friction material development, on-road brake test is the final test normally performed to evaluate and validate the formulation as the brake friction material is actually tested under its real life application conditions. A brake test is basically a deceleration test carried out between two speeds. Data taken during the test is used to calculate the time taken, distance travelled and deceleration. A few other additional parameters such as brake hydraulic pressure, brake pedal effort and temperature of the friction materials would also be normally measured.

The developed formulations were subjected to on-road test as per ECE R13 and shall achieve a minimum mean fully developed deceleration (MFDD) requirements as shown in Table 11. Alternatively, the braking performance may also be evaluated in terms of the stopping distance. The tests consists of three (3) test modes (cold, fade and recovery) simulating real conditions of brake lining material temperature during its service. Figure 13 shows a typical display of Dewetron DEWE-5000 acquisition system during fade and recovery tests.

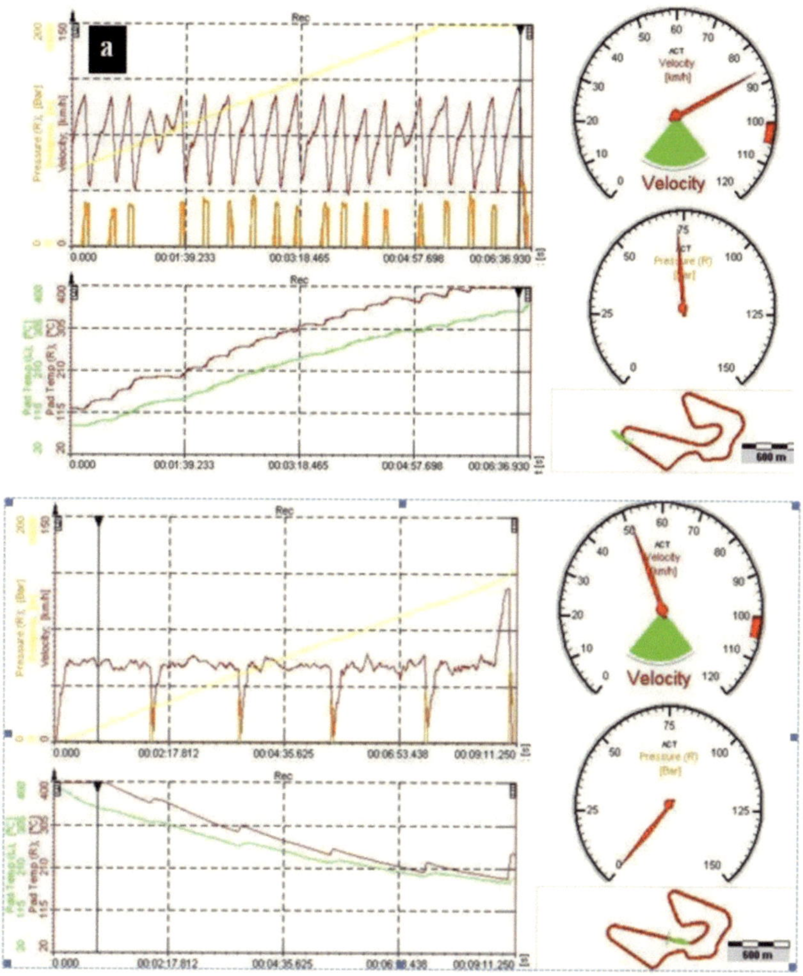

Fig. 13. Display of data acquisition system; (a) fade test, (b) recovery test

Tests	Mean fully developed deceleration (m/s^2)
cold effective	6.43
heat fade	75 % of that prescribed and 60 % of figure recorded in the cold effectiveness test
recovery	not less than 70 %, nor more 150 %, of figure recorded in the cold effectiveness test
Pedal force	Shall be more than 500N

Table 11. Minimum requirements of the performance tests (ECE R13)

Test results of on-road braking performance for all the developed friction material formulation and a commercial pad are shown in Table 12 Figure 14. Out of the 10 prototype and 1 commercial sample tested for on-road performance, four samples do not fully comply with ECE's requirements, namely, sample SM2, SM4, SM7 and SM9. Samples SM2, SM7 and SM9 require pedal forces of 536N, 646N and 980 N, respectively under cold test conditions, which are exceeding the maximum permitted pedal force of 500 N as shown in Table 11. As such, further tests (i.e. fade and recovery) were not performed on these samples and the samples were eliminated. Higher pedal force requires more driver effort to stop the vehicle, which may stress the leg, especially for the lady driver. Sample SM4 does comply with deceleration requirement under fade test with MFDD of 4.18 m/s^2 which is less than the required value of 4.81 m/s^2 (i.e. >75% of 6.43). Test results also show that all other samples (COM, SM1, SM3, SM5, SM6, SM8, SM10) fully comply with cold-, fade- and recovery-tests requirements. Sample SM3, SM5, SM6 and SM8, though, shows higher values of MFDD during recovery tests than the cold effectiveness tests, which is allowed by this regulation which states that the MFDD can go up to 150% of the figure recorded in cold effective test.

Sample	Test Mode	Pedal Force [N]	MFDD [m/s2]	Pad Temp. (deg C)	
				Left	Right
COM	Cold	138	7.65	93	126
	Fade	150	5.49	350	456
	Recovery	133	7.50	184	242
SM1	Cold	254	8.00	161	172
	Fade	230	8.10	212	221
	Recovery	218	8.24	158	140
SM2	Cold	536	8.14	169	177
	Fade	Not performed, F>500 N for Cold Test			
	Recovery	Not performed, F>500 N for Cold Test			
SM3	Cold	114	7.33	103	158
	Fade	120	5.77	394	459
	Recovery	116	7.39	202	260
SM4	Cold	96	6.83	115	110
	Fade	94	4.18	416	312
	Recovery	98	4.95	181	155
SM5	Cold	129	7.46	93	89
	Fade	132	5.73	253	273

Sample	Test Mode	Pedal Force [N]	MFDD [m/s2]	Pad Temp. (deg C)	
				Left	Right
	Recovery	136	7.67	139	137
SM6	Cold	128	8.1	155	154
	Fade	130	7.07	272	260
	Recovery	134	8.38	143	152
SM7	Cold	646	6.02	202	193
	Fade	Not performed, F>500 N for Cold Test			
	Recovery	Not performed, F>500 N for Cold Test			
SM8	Cold	133	7.29	111	99
	Fade	135	5.67	345	289
	Recovery	132	7.32	169	143
SM9	Cold	980	-	292	320
	Fade	Not performed, F>500 N for Cold Test			
	Recovery	Not performed, F>500 N for Cold Test			
SM10	Cold	104	7.05	316	277
	Fade	109	5.72	515	527
	Recovery	117	5.77	333	332

Table 12. On-road performance test results

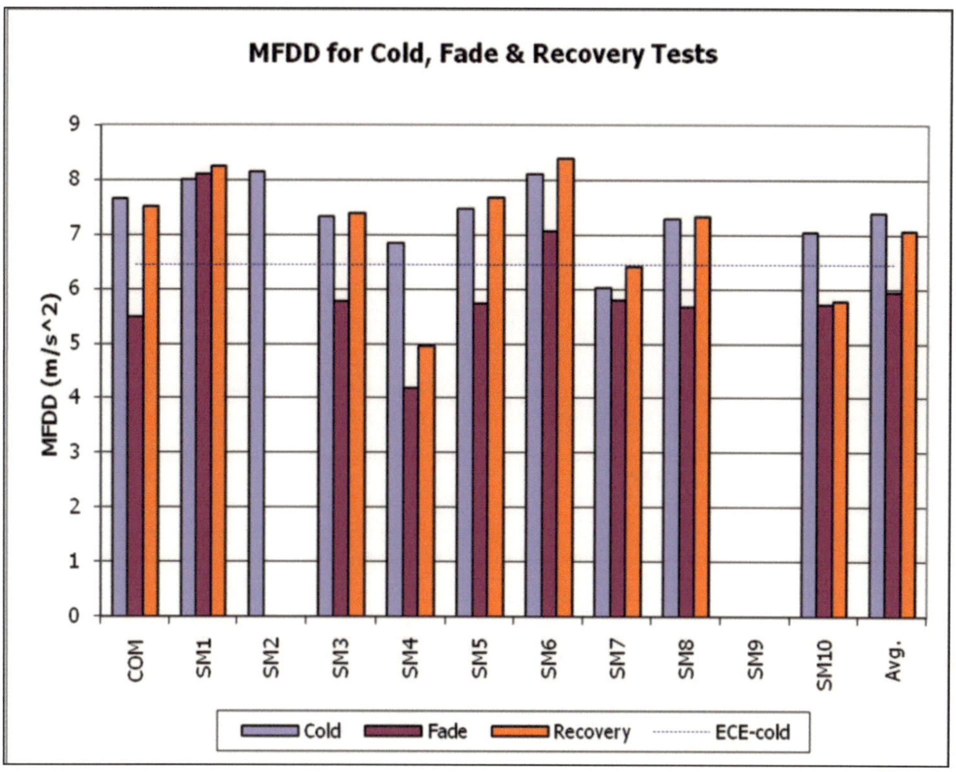

Fig. 14. MFDD for Cold, Fade and Recovery Tests

In the braking process, kinetic energy of a moving vehicle is converted into thermal energy. The generation of heat is due to the friction between the friction materials and brake disc. The heat generated is dissipated to the surroundings by the brake disc and friction materials. The ability of brake disc to dissipate this heat significantly affects the performance and wear life of the friction materials. Heat generated during braking result a phenomenon known as heat fade where the friction fall at elevated temperature. This fading effect is associated with the decomposition of the organic compound. The decomposition of the binder takes place between 250 and 475 °C (Ramousse et al. 2001). This phenomenon results in a reduction of friction as the temperature increased as observed by Rhee 1971 and Talib et al. 2001. This sudden drop of friction results in lower brake performance, in which longer braking distance is required before the moving vehicle can be stopped.

Figure 15 shows that there is no direct correlation between thick loss of prototype brake pad during brake dynamometer and on-road test. The test sequences and braking parameters of the brake dynamometer and on-road tests are not the same. For homogeneous materials, the thickness loss of the two test methods will be producing the same tend. But for friction materials, this postulation does not apply. This was taught due to heterogeneous properties of friction materials, where wear of friction materials is dependent on the mechanical, chemical, thermal properties as well as the microstructure. On-road braking test results give a better picture of the performance of the developed friction material formulations in real life applications as compared with the laboratory test data.

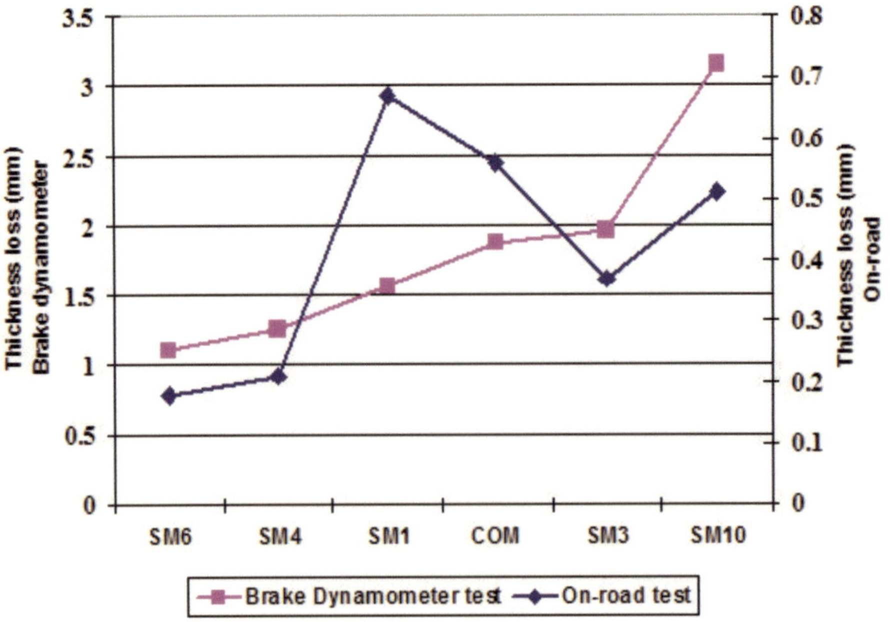

Fig. 15. Thickness loss of brake pad during brake dynamometer and on-road test

3.4 Microstructural

Figure 16, shows the photograph of sample SM6 after subjected to on-road performance test. The pad has pitting, grooving and moderate resin bleed. It was observed there is no flaking or surface cracking evident on the worn surface of brake pad. The surface of rotor has light lining transfer and light grooving.

Fig. 16. Post on-road performance test photograph on brake pad and brake disc

The worn surface after on-road performance test were analyzed using scanning electron microscope equipped with energy-dispersive X-ray analyzer (EDX). Microstructureal examaination on worn surface revealed that the mechanism composed a complex mixture of abrasion, adhesion and delamination as shown in Figure 17. Figure 17a show a manifestation of abrasion wear mechanisms where the harder peak asperities were ploughed into the surface. Figure 17b and c show a manifestation of adhesion mechanism. Adhesion wear mechanism composed a process of two-way transfer during sliding caused the formation of transfer layers on both sides of the sliding surfaces (Figure 17b) as observed elsewhere (Kerridge & Lancaster 1956; Chen and Rigney 1985; Talib et al. 2003) due to mechanical alloying as reported by Chen at el. (1984). Figure 17c shows transfer layers appeared to be sheared and flattened and smeared on their surfaces during then raking process. Thus it can be concluded that the wear surfaces became smoother with increase in braking time. Figure 17d shows a sympton of delamination mechanism where it reveled the wear particles flake off from the wear surface when reaching the critical length.

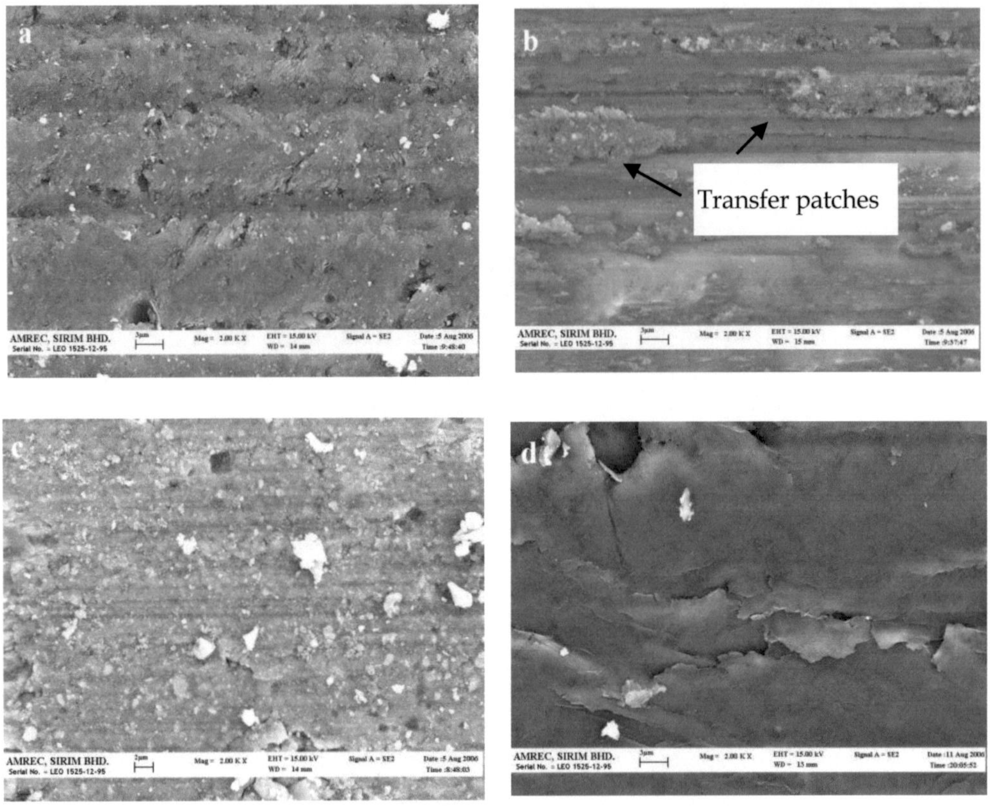

Fig. 17. Wear mechanism observed during braking process; (a) abrasion, (b) adhesion - generation of transfer patches, (c) adhesion - smearing and (d) delamination

4. Conclusions

Characteristic of friction materials is very complex to predict and it is a critical factor in brake system design and performance. To achive ideal brake friction material characteristic such as constant a constant coefficient of friction under various operating conditions, resistance to heat, water and oil fade, low wear rate, posses durability, heat stability, exhibits low noise, and not to damage brake disc, some requirements have to be compromised in order to achieve some other requirements. This can be done by changing the type and weight percentage of the ingredients in the formulation. This works shows

The following phenomena could be postulated based on the physical and mechanical, friction and wear test results using CHASE machine, brake effective test results using brake inertia dynamometer, and braking performance test results;

i. Test results show that there is no simple relationship between the physical and mechanical properties and thus, these test results could not be used to screen the developed samples. The physical and mechanical properties are used for quality control in producing friction material with consistent physical and mechanical properties.

ii. Friction and wear characteristics obtained using CHASE machine could not be simply related to physical and mechanical characteristics.
iii. Friction and wears assessment tests using CHASE brake lining machine can be used for screening of friction material formulations during development as well as for quality control. Thickness loss using CHASE machine cannot be used to predict thickness loss using brake inertia dynamometer.
iv. The test sequence and parameters of brake dynamometer cannot simulate exactly all the braking pameters and enviroment of on-road test condition. Thus there is no simple correlation between the brake dynamometer test results with on-road performance results.
v. The final selection of the best formulation is based on on-road performance test results. However, the prototype samples need to be subjected to endurance tests to ensure that formulations can perform under real life application conditions.

The development and validation of a friction material involve a significant amount of testing in laboratory and on the road. On-road brake test is the final test normally performed to evaluate and validate the formulation as the brake friction material is actually tested under its real life application conditions. Thus, vehicle testing on the test track is the ultimate measure for the overall assessment of the brake performance testing and evaluation. Out of the ten (10) developed friction formulations, only 6 samples complied with the on-road braking performance requirements. However, further investigations on the performance and wear of the developed brake pads need to be conducted on actual intended application on various real road conditions.

During braking process, the brake pad is pressed against a rotating disc which in turn slows down the rotation of the wheels of a vehicle and thus stops the vehicle. In the process of decelerating a moving vehicle, kinetic energy is converted into thermal energy. This accumulated heat is absorbed by the brake pads and brake disc before being dissipated to the atmosphere. The accumulation of heat causes high surface temperatures in the lining materials and the brake disc which leads to the changes to the mechanical, chemical and wear mechanism. The brake lining materials wear off as a result of friction between the lining materials and the brake disc. Micro-structural changes on the worn surface of the brake reveals that the wear mechanisms operated during braking include adhesion, abrasion and delamination. The wear mechanisms operated during braking are rather complex with no single mechanism was found to be operating fully.

5. Acknowledgment

This research was supported by Ministry of Science, Technology and Innovation, Malaysia by providing research grant and SIRIM Berhad for providing research facilities.

6. References

Begelinger, A. and Gee, A.W.J. 1973. A New Method for Testing Brake Lining Material. *ASTM Special technical publication 567*, Philadelphia: American Standard for Testing and Materials. Pp. 316-334.

Bros, J. & Sciesczka, S.F. (1977). The investigation of factors influencing dry friction in brakes. *Wear,* Vol 34, pp. 13139.

Chen, L.H. & Rigney, D.A. 1985. Transfer during unlubricated sliding of selected metal systems. *Wear,* Vol. 59, pp. 213-221

Cho, M. H.; Kim, S. J., Kim, D. & Jang, Ho. (2005), Effects of ingredients on tribological characteristics of a brake lining: an experimental case study, *Wear,* Vol. 258, pp. 1682–1687

ECE Regulation R 13. Uniform Provisions Concerning the Approval of Passenger Cars with Regard to Braking

Filip, P., Kovarik, L, Wright, M, A. 1995. *Automotive brake lining chracterization.* Proceeding of the 8th international pacific conference on automobile engineering, 4-9 November, pp. 417-422. Yokohama: Society of Automobile Engineers of Japan, pp. 34-422

Hsu, S.M.; Shen, M.C. & Ruff, A.W. (1997). Wear prediction for metals. *Tribology International* 30, pp. 377-383.

Ingo, G.M.: D'Uffizi, M., Falso, G., Bultrini, G., & Padeletti, G. (2004), Thermal and microchemical investigation of automotive brake pad wear residues, *Thermocimica Acta,* Vol. 418, pp. 61-68.

Jacko, M.G. (1977). Physical and chemical changes of organic disc pad. In: *Wear of materials,* Glaeser, W.A., Ludema, K.C. & Rhee, S.K. (Ed). pp. 541-546. ASME, New York

Jacko, M.G.; Tang, P.H.S. & Rhee, S.K. (1984) . Automotive friction materials evaluation during the past decade. *Wear,* Vol. 100, pp. 503 - 515

Jang, H.; Ko, K., Kim, S.J., Basch, R.H. & Fash, J.W. (2004). The effect of metal fibers on the friction performance of automotive brake friction materials, *Wear,* Vol. 256, pp. 406–414

JIS D 4418: 1996. Test Procedure of Porosity for Brake Linings and Pads of Automobiles

JIS D 4421: 1996. Method of Hardness Test for Brake Linings, Pads and Clutch facings of Automobiles

Lim, S.C., Ashby, M.F. & Bruton, J.H. 1987. Wear-rate transitions and their relationship to wear mechanisms. *Acta Metall.,* Vol. 35, No. 6, pp. 1343 - 1348.

Lu, Y. (2006). A combinatorial approach for automotive friction materials: Effects of ingredients on friction performance, *Composites Science and Technology,* Vol. 66, pp. 591–598

Mokhtar, M.O.A. 1982. The effect of hardness on the frictional behaviour of metals. *Wear,* Vol. 78, pp. 297- 305.

More, A.J.W. & Tegart, W.J.McG. 1952. Relation between friction and hardness. *Proc. Royal Soc. A,* Vol. 212, pp. 452-458.

MS 474: PART 1: 2003. Methods of Test for Automotive Friction materials (Brake Linings, Disc Pads and Bonded Shoe): Part 1: Specific gravity (First Revision)

Mutlu, I.; Eldogan, O. & Findik, F. (2006), Tribological properties of some phenolic composites suggested for automotive brakes, *Tribology International,* Vol. 39, pp. 317–325

Österle, W. & Urban, I. (2004), Friction layers and friction films on PMC brake pads, *Wear,* Vol. 257, pp. 215-226.

Rhee, S.K. (1973). Wear mechanism at low temperature for metal reinforced phenolic resins. *Wear*, Vol. 2, pp. 261-263.

Rhee, S.K. (1976). High Temperature wear of asbestos reinforced friction materials. *Wear*, Vol. 37, pp. 291-297.

Rhee, S.K. 1971. Wear of Material - Reinforced Phenolic Resins. *Wear*, Vol. 18, pp. 471-477.

Rigney, D. A. (1997). Comments on sliding of metals. *Tribology international*, Vol. 30, No. 5, pp. 361-367.

SAE 661: FEB 97. Brake Lining Quality Test Procedure, Society of Automotive Engineers, Warrendale, Pennsylvania.

SAE Recommended Practice J866 list for codes and associated friction coefficient , Society of Automotive Engineers, Warrendale, Pennsylvania. Society Automotive of Enginner

SAE J2522: 2002. Dynamometer Global Brake Effectiveness Recommended Practice, Society of Automotive Engineers, Warrendale, Pennsylvania

Sanders, P.G., Dalka, T.M., and Basch, R.H. (2001). A reduced-scale brake dynamometer for friction characterization, *Tribology International*, Vol. 34, pp. 609-615.

Sasaki, Y. 1995. *Development philosophy of friction materials for automobile disc brakes.* The eight international pacific conference on automobile engineering. 4-9 November, hlm. 407-412. Yokohama: Society of Automobile Engineers of Japan.

Scieszka, S.F. 1980. Tribological phenomena in steel-composite brake material friction pairs. *Wear*,Vol. 64, pp. 367 - 378.

So, H. 1996. Characteristics of Wear Results Tested by Pin-on Disc at Moderate to High Speeds. *Tribology International*, Vol. 29, No. 5, pp. 415 - 423

Talib, R.J. (2001) Investigation Into The Surface and Bulk Wear Morphology of Automotive Friction Materials. PhD Thesis, National University of Malaysia, Bangi, Malaysia.

Talib, R. J.; Muchtar, A. & Azhari C.H. (2007). The Performance of Semi-Metallic Friction Materials for Passenger Cars, *Jurnal Teknologi*, Vol. 47(A), pp. 53-72

Talib, R.J.; Muchtar, A. & Azhari, C.H. (2003). Microstructural characteristics on the surface and subsurface of semi-metallic automotive friction materials during braking process, *Journal of Material Processing Technology*, Vol. 140, pp. 694-699.

Talib, R.J.; Shaari, M.S., Ibrahim, W.M.A.W., Kemin ,S. & Kasiran, R. (2006). Properties Enhancement of Indigenously Developed Brake Pad for Light Rail Transit, In: *Brake Friction Materials*, Darius G. S. (Ed). Pp. 79-86, Shah Alam, ISBN 983-3644-72-4, UPENA, Malaysia

Talib, R.J.; Azimah, M.A.B., Yuslina, J., Arif S.M. & Ramlan K. (2008), Analysis on the Hardness Characteristics of Semi-metallic Friction Materials. *Journal Solid State Science & Technology*, Vol. 16, No. 1, pp. 124-129

Tanaka, K.; Ueda, S. & Noguchi, N. (1973) . Fundamental studies on the brake friction of resin-based friction materials. *Wear*, Vol. 23, pp. 349-365.

Todorovic, J. (1987) . Modelling of the tribological properties of friction materials used in motor vehicles brakes. *Proc. Instn. Mech. Engrs.* Vol. C 226, pp. 911-916.

Tsang, P.H.S.; Jacko, M.G., and Rhee, S.K. (1985). Comparison of CHASE and Inertial brake dynamometer testing of automotive friction materials, *Wear*, Vol. 103, pp. 217-232.

Zhigao, X. and Xiaofei, L. 1991. A Research for the Friction and Wear Properties of a Metal-fiber-reinforced Composite Material. *In Mechanical Properties Materials Design. International Symposia Proceedings* 1991. Boqun Wu (ed.). Amsterdam: Elsevier Science Publisher. pp. 611-615.

2

Porous Metals and Metal Foams Made from Powders

Andrew Kennedy
Manufacturing Division, University of Nottingham, Nottingham, UK

1. Introduction

Porous materials are found in natural structures such as wood, bone, coral, cork and sponge and are synonymous with strong and lightweight structures. It is not surprising that man-made porous materials have followed and those made from polymers and ceramics have been widely exploited. The commercialisation of porous metals has lagged somewhat behind, but has received a boost following a surge in worldwide research and development in the early 1990's.

The unique combination of physical and mechanical properties offered by porous metals, combinations that cannot be obtained with dense metals, or either dense or porous polymers and ceramics, makes them attractive materials for exploitation. Interest mainly focuses on exploiting their ability to be incorporated into strong, stiff lightweight structures, particularly those involving Al foams as the "filling" in sandwich panels, their ability to absorb energy, vibration and sound and their resilience at high temperature coupled with good thermal conductivity.

The applications for porous metals (metals having a large volume of porosity, typically 75-95%) and metal foams (metals with pores deliberately integrated into their structure through a foaming process) depend on their structure. Closed cell foams, which have gas-filled pores separated from each other by metal cell walls, have good strength and are mainly used for structural applications. Open cell foams, which contain a continuous network of metallic struts and the enclosed pores in each strut frame are connected (in most cases these materials are actually porous or cellular metals), are weaker and are mainly used in functional applications where the continuous nature of the porosity is exploited. Examples of structures for both these types of porous metals are shown in Figure 1.

Table 1 summarises the potential uses for porous metals and metal foams, highlighting the relevant attributes that make them suitable for that particular application (Ashby et al., 2000). Specific examples of current applications for porous metals and metal foams can be found in (Banhart, 2001), but it should be remarked that this is a dynamic area and new applications and components are continually being developed.

This short review describes the main powder metallurgy-based manufacturing routes, highlighting the key aspects of the relevant technology involved and the types of foam structures that result.

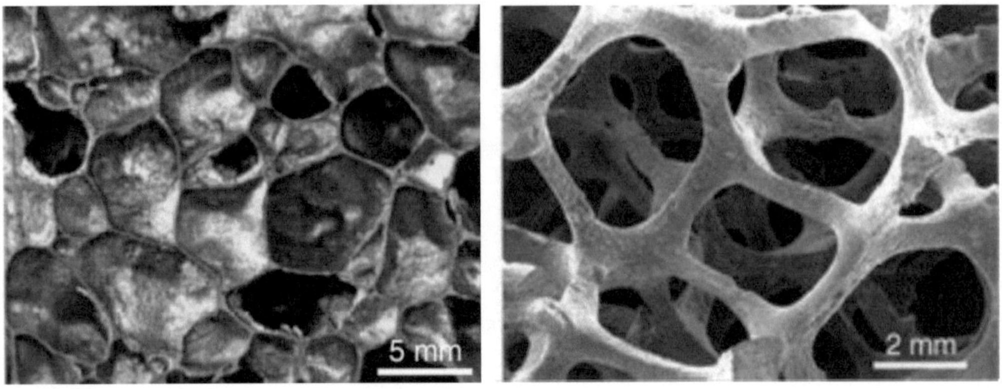

Fig. 1. Micrographs of (left) a closed cell foam and (right) an open cell foam, or more correctly, a cellular metal (Zhou, 2006).

Application	Relevant Attributes
Lightweight structures	Excellent stiffness-to-weight ratio when loaded in bending
Mechanical damping	Damping capacity is larger than solid metals by up to 10x
Vibration control	Foamed panels have higher natural flexural vibration frequencies than solid sheet of the same mass per unit area
Acoustic absorption	Open cell metal foams have sound-absorbing capacity
Energy absorbers / packaging	Exceptional ability to absorb energy at almost constant pressure
Heat exchangers	Open-cell foams have large accessible surface area and high cell-wall conduction giving exceptional heat transfer ability
Biocompatible inserts	Cellular texture stimulates cell growth
Filters	Open-cell foams for high-temperature gas and fluid filtration

Table 1. Potential application areas for porous metals and metal foams adapted from (Ashby et al., 2000).

2. Processing methods for metal foams

There are many different ways to produce porous metals and metallic foams and these methods are usually classified into four different types of production, using liquid metals, powdered metals, metal vapour or metal ions. The use of powdered metals as the starting material for foam production offers the same types of advantages (and often the same limitations) as conventional powder metallurgical processes. If a particular metal or alloy can be pressed and sintered, there is a high likelihood that it can be made into a porous metal or metal foam.

2.1 Porous metals produced by powder sintering

2.1.1 Pressureless sintering

Loose pack, pressureless or gravity sintered metal powders were the first form of porous metals and are still widely used as filters and as self-lubricating bearings. The porosity in these components is simply derived from the incomplete space filling of powders poured into and sintered in a die. With packing densities broadly in the range of 40-60%, but affected by particle shape, size and vibration, the porosities in these structures are well below those for most porous metals. The simplicity of the process means that porosity can be included in a wide range of metals, limited only by the ability to sinter the metal in an appropriate die. The process is most commonly used to sinter bronze powders to make bearings, an example of which is shown in Figure 2, but porous structures from titanium, superalloys and stainless steel have also been made in this way.

Fig. 2. Porous bronze made by pressureless sintering of approximately 100 μm diameter powders [Eisenmann, 1998].

In the case of Al alloys, where sintering is made difficult by the surface oxide layer covering the powder particles, milling of the powder with sintering aids such as Sn and Mg is required. The limitations imposed by the need to sinter in a die, principally on the size of the product and the productivity, can be mitigated by performing die (or other) compaction processes to increase the green strength of the compact so that it is sufficient to perform containerless sintering. The inevitable consolidation that is involved will, of course, decrease the already low porosity.

In an effort to increase the porosity in these parts, powders have been replaced by metal fibres, made by processes such as melt spinning, which with higher aspect ratios, exhibit lower packing densities, making them suitable for a wider range of applications (Anderson and Stephani, 1999). A further development of this, leading to much higher porosities, is the sintering of hollow metal spheres (which themselves are made via a powder route - to be described later). Spheres with diameters ranging from 1.5 to 10 mm, with wall thicknesses from 20 to 500 μm can be arranged to produce both open and closed pore structures. Open structures can be obtained in the same way as for powders and compaction can be used to deform the spheres to polyhedral bodies, reducing the degree of open porosity. True closed pore structures can be obtained by filling the interstices between the spheres with metal powder followed by a sintering treatment. Porosity is contained both within and between the hollow spheres and porosities in the range of 80-97% are reported (Anderson et al., 2000). Examples of an open cell sintered hollow sphere structure and the spheres in cross section, showing their thin walls, are presented in Figure 3.

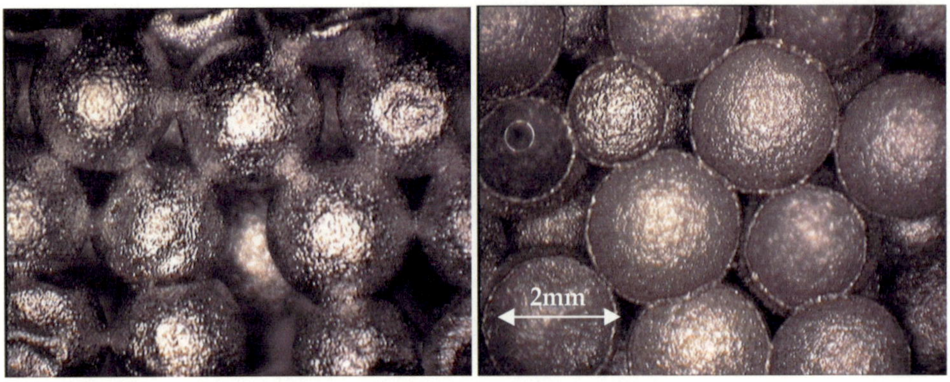

Fig. 3. Steel hollow spheres (left) sintered to form an open cell foam and (right) sectioned.

The advantage of porous structures made from sintered powders or hollow spheres is that there is good control of the volume fraction and to a lesser extent, the geometry and size of the pores, leading to reproducible structures and properties. Sintered metal powder and fibre structures are mainly suited to applications based on filtration, catalysis or heat exchange. The low density structures that can be made from sintered hollow spheres can be used for lightweight structural parts and for energy and sound absorption.

2.1.2 Gas entrapment

Internal porosity can also be developed in metal structures by a gas expansion (or foaming) process based on hot isostatic pressing (HIPing) (Martin and Lederich, 1992). Initially, following the standard method for HIPing of metal powders, a gas-tight metal can is filled with powder and evacuated. In a deviation from common practise, the can is then filled with argon gas at pressures between 3 and 5 bar before being sealed, isostatically pressed at high temperature and then worked to form a shaped product, normally a sheet. Porosity is generated by annealing the part. When holding at elevated temperature, the pressurised argon gas present within small pores in the structure causes the material to expand (foam) by creep. As HIPing is an effective method for sintering and the can material can be made from the same material as the powder, this process could be used for many different metal powders. The use of a can means that sandwich-type structures, consisting of a lightweight foam core and thin, solid face sheets are produced. These types of structures are ideal for lightweight construction.

Porous bodies with typically 20–40% of isolated porosity are obtained and theoretical considerations show that no more than 50% porosity can be expected (Elzey and Wadley, 2001). Figure 4 shows a schematic representation of the process, which has been used to make porous titanium sandwich structures for the aerospace industry, without the need for complex joining methods. Disadvantages include low porosity and irregular-shaped pores.

2.1.3 Reactive processing

In contrast to the gas entrapment method, porosity is evolved much more rapidly when foaming occurs in highly reactive multi-component powder systems such as those which undergo self-propagating high temperature synthesis (SHS). The highly exothermic

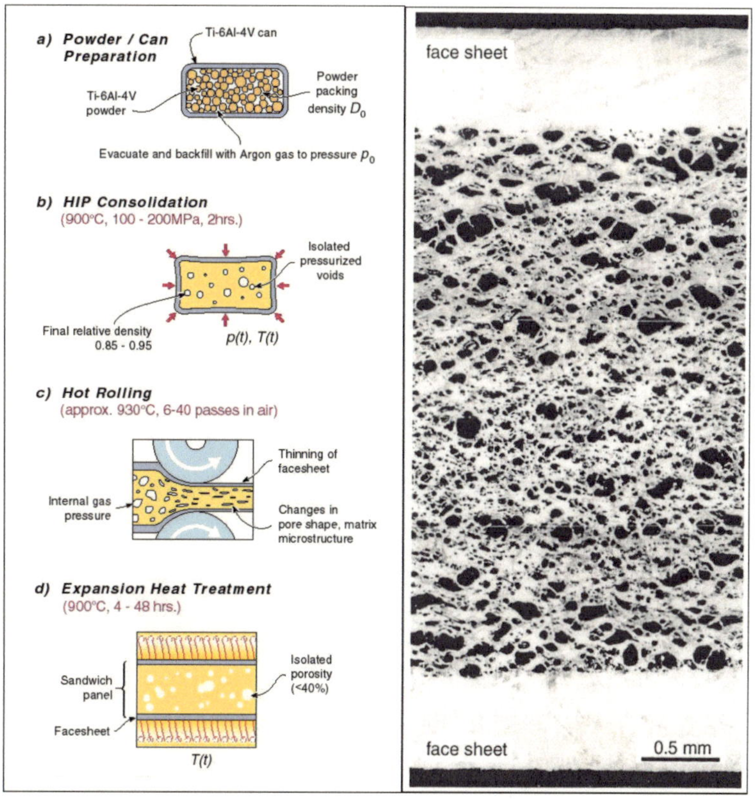

Fig. 4. A schematic (left) of the processing steps used to manufacture titanium alloy foam sandwich panels by gas entrapment (Ashby et al., 2000) and (right) the morphology of a TiAl6V4 sandwich structure (Banhart, 2001).

reactions, initiated either by local or global heating of compacted powder mixtures to the reaction ignition temperature, lead to vapourisation of hydrated oxides on the powder surfaces and the release of gases dissolved in the powder. The reacting powder mixture heats up rapidly to form a liquid containing (mostly hydrogen) gas bubbles and when the reaction is complete, cools rapidly, entrapping the gas to form a foam. A schematic of this process is shown in Figure 5 (Kanetake and Kobashi, 2006).

Gas formation and foam expansion can be augmented by the addition of vapour forming phases such as carbon (which burns in air to produce CO) or foaming agents which react together to increase the reaction temperature and produce fine particles that stabilise the foam. As foaming takes place in the liquid state, stabilisation of the bubbles is needed to avoid rapid collapse of the foam structure. Figure 5 shows how a reactive Ti+B_4C foaming agent increases the porosity in a Ni-Al powder mixture from 30 to 90% by 5% addition. It can be seen that the pores are irregular in shape, as is the shape of the expanded foam. Although the process is relatively simple, the production of foams is limited to combinations of materials that react exothermically, some metal-metal systems but typically metals and carbon or carbides, or metals and oxides. These limitations, coupled with the difficulty controlling the expansion process and defining the shape of the expanded foam, mean that few, if any, commercial foam products are made this way.

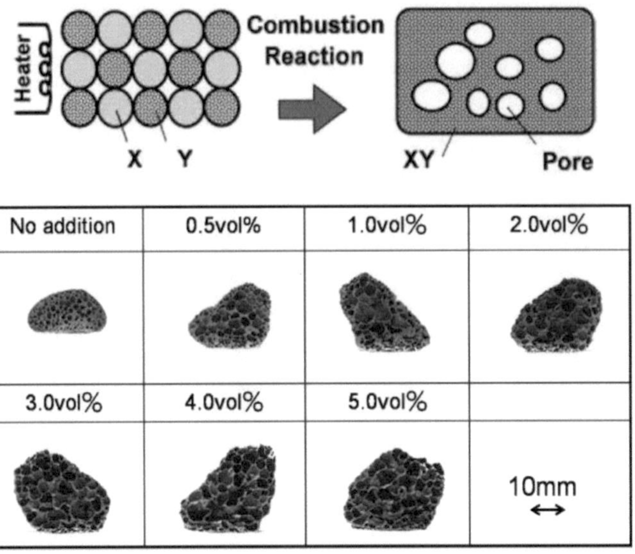

Fig. 5. A schematic (top) of the reactive powder process used to make metal foams and (bottom) cross-sections of $NiAl_3$ foams containing different addition levels of a Ti + B_4C foaming agent mixture, after (Kanetake and Kobashi, 2006).

2.1.4 The addition of space-holding fillers

Many of the processes already mentioned have shortcomings in either the rather limited levels of porosity that are achievable, or in the formation of pores that are highly irregular with a wide distribution in sizes. A simple development of standard PM practices, but incorporating a volume of sacrificial space fillers, offers a solution to both these problems.

Porous metals are produced by mixing and compacting metal powders with a space holder which is later removed either during or after sintering, by dissolution or thermal degradation, to leave porosity (Zhao and Sun, 2001). A schematic of this process is shown in Figure 6. This simple method has the advantage that the morphologies of the pores and their size are determined by the characteristics of the space holder particles and the foam porosity can be easily controlled by varying the metal/space holder volume ratio. Addition levels of the space holder are typically between and 50% and 85%. These are sufficiently high that they are interconnected and hence can be removed easily. Above 85% the structure of the struts is unlikely to be continuous and below 50%, residual space filler will be enclosed within the structure, making removal very difficult.

Commonly, space fillers take the form of polymer granules and water soluble salts, but can also be metal powders and ceramic or polymer hollow spheres (if the hollow spheres aren't removed then these materials are, strictly speaking, syntactic foams). Two distinct approaches may be taken. In the first, the space filler can either thermally decompose, sublime or evaporate below the sintering temperature of the metal matrix. This requires the resulting skeletal metal structure to have good strength, which is affected during the compaction stage. Polymer powders (for example PMMA), carbamide granules or Mg grains are commonly used. In the second approach, the space filler has a higher melting point than the sintering temperature and is removed, after sintering, by dissolution in a

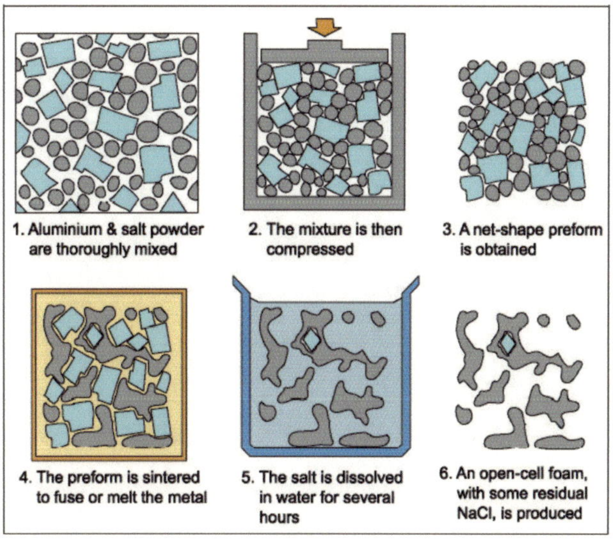

Fig. 6. Production of an open cell foam by sintering a mixture of metal powder and a removable agent (Ashby et al., 2000).

suitable solvent in which the metal matrix should be unaffected, most commonly water. In the case of Al alloys, NaCl is the favoured space holder as it melts at approximately 800°C and can be dissolved fairly rapidly in warm water. For higher melting point metals, $NaAlO_2$, with a melting point of 1800°C, is used.

In addition to the decomposition or melting temperature and general ease of processing, including solubility in the solvent, space fillers are also selected based on their inertness and lack of solubility in the matrix, as well as the availability in the size and shape desired (typically in the range of hundreds of μm to a few mm and often spherical space fillers are preferred). As salts (such as NaCl) are brittle materials and, particularly when ground, are angular in shape, melting and granulation methods have been used to produce them in spherical form (Goodall and Mortensen, 2007). If these granules are porous, this has the additional benefit of more rapid removal from the sintered part due to the granule being able to disintegrate as well as dissolve. Figure 7 shows porous beads made by the controlled agglomeration of fine salt particles and the structure of the resulting porous metal (Jinnapat and Kennedy, 2010). The cell structure is clearly interconnected via windows between neighbouring pores, the number and size of which are dependent upon the co-ordination number and contact area for the packed spheres.

Processing difficulties can arise during mixing the metal powder with the often much larger space holders. Segregation results in defects in the final product, usually in the form of incomplete struts and cells. Liquid binders are used to promote homogeneous mixing, often by ensuring that the space filler is coated with the metal powder. An alternative approach is to vibrate the finer metal powders into the interstices within a packed bed of the larger space holder particles. The metal powder - space holder mixtures are then compressed to form a compact with sufficient strength to be sintered free-standing and, in the case of Al to disrupt surface oxide films so that sintering can take place. Compaction processes can include all of those used in standard metallurgical practices, including metal injection moulding.

Fig. 7. Images showing (left) porous spherical salt beads and (right) the microstructure of the resulting cellular stainless steel (Jinnapat and Kennedy, 2010).

Despite a few drawbacks, for example if space fillers such as NaCl are not completely removed this can lead to corrosion and the size of parts produced is rather limited, in part due to slow dissolution of the space fillers which might take days even for small parts, this method is a favoured route for the manufacture of porous metals from a wide range of materials. It is particularly suited to those metals with high melting points and is a common route for the production of porous biomedical devices made from Ti or NiTi.

2.1.5 Additive manufacturing

Porous metal structures can be built, layer upon layer, using processes such as selective laser sintering (SLS), direct metal typing or 3D direct metal printing. 3D parts are constructed by stacking these layers, the geometry of which is defined by a CAD model.

Direct metal laser sintering uses a high power laser to sinter metal powders on the surface of a powder bed. After each cross-section is scanned, the powder bed is lowered by one layer thickness, a new layer of material is applied on top, and the process is repeated until the part is completed. Two-component powders can be used, comprising metals coated with a polymer, where the laser only melts the coating. Products made in this way, however, still need a secondary sintering step to produce sufficiently robust parts. Steel and titanium foams can be made by both these methods, although with polymeric binders, contamination of Ti with carbon occurs and direct metal laser sintering is, therefore, preferred.

Direct metal printing works in a similar way, spraying a polymeric binder, which is dispensed through a print head, over a powder bed. 3D metal typing or screen printing uses a metal powder mixed with a binder which is then spread over patterned masks and cured layer by layer (Andersen et al., 2004). For both these processes, subsequent polymer-removal and sintering steps are required. Figure 8 shows cellular structures made from 316L stainless steel by direct typing. Among the benefits of these processes are high levels of design flexibility, including small, precise cellular structures with complex internal geometries.

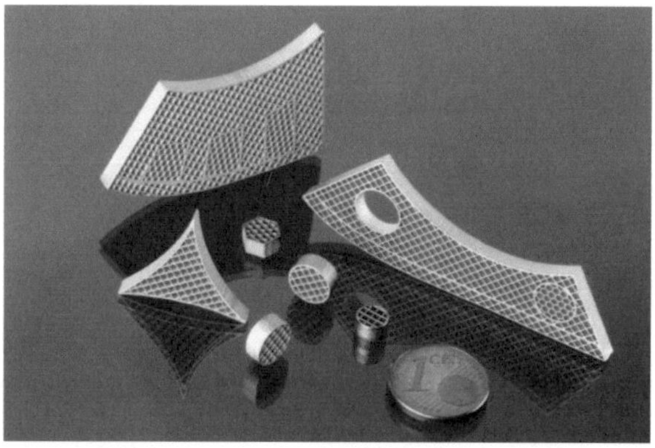

Fig. 8. Cellular structures made from 316L stainless steel by direct typing (Andersen et al., 2004).

2.2 Metal powder slurry processing

2.2.1 Metal powders slurries

Slurry processing of ceramics has been used for many years to produce bulk products, coatings, films and foams. The wealth of scientific research and resulting literature pertaining to this area is not mirrored for metal powders slurries, the need for processing in this way is perhaps not so great given the many alternative forming methods for metals. The production of stable (non-agglomerated, non-sedimenting) slurries containing metal powders provides significant challenges. Metal powders are larger in size than those used for ceramic processing and have a higher density, making them more difficult to keep suspended. Metal powder slurries, like their ceramic counterparts, are usually based on aqueous systems to which a suspending agent (to increase the viscosity) and dispersants (to prevent particle flocculation) are added (Kennedy and Lin, 2011). Other additions may be needed to lower the surface tension, change the pH or facilitate gelation.

The advantages of using metal slurries as the basis for producing foams is that constraints imposed by poor compressibility are removed, shaping can be performed by gelling in simple rubber moulds and gas bubbles can be introduced into the system by whisking or through the addition of gas-forming agents (although this means that the foam has to be stabilised). Although not well established, there are likely to be limitations to the materials that can be used based on interactions between the metal and the solvent (corrosion or reaction) and the ability to form a stable slurry.

2.2.2 Slurry coating of polyurethane foams

Porous metals can be produced by a replication method using an open cell polyurethane (PU) foam or sponge as a template (Quadbeck et al., 2007). In this process the polymer foam is first coated with a metal powder slurry, usually performed by immersion or by spraying. Excess slurry is removed by squeezing the foam, often by passing it through rollers. Without this, the cells may become partially closed due to the formation of liquid films bridging the cell struts. After coating and drying, the template is removed by thermal degradation and the resulting, fragile metal skeleton is further heated to sinter the metal powder particles, forming a rigid cellular metal structure.

Figure 9 shows a sintered porous stainless steel structure alongside the polymer foam used in the replication process. The strut structure is also shown and it is clear from the cross section that the struts are hollow (due to the space vacated by the polymer) and the walls of the struts are porous. Open cell structures with cell sizes between 10–80 pores per inch (about 2.5-0.2 mm) can be produced with precise pore structures with total porosities as high as 96%. This total porosity also includes the porosity in the sintered metal struts, but open cell porosities as high as 90% can be achieved.

This type of approach can be used to make metallic hollow spheres (which are themselves sintered to produce cellular metals as was described earlier). To make hollow spheres, expanded polystyrene spheres are coated with a metal powder slurry and then sintered, during which the polystyrene is removed and the metal forms a dense metal shell (Andersen et al., 2000).

Fig. 9. Images (left) showing a PU foam and a stainless steel foam made by slurry coating and (right) the microstructure of the struts showing, inset, their hollow nature.

2.2.3 Slip reaction foam sintering

Metal powder slurries can also be foamed by the insitu generation of a gas. In the slip reaction foam sintering (SRFS) process, the slurry (or slip) contains additives which stabilize the slip during processing (Angel et al., 2004) and to this a solution of ortho-phosphoric acid in either water or alcohol is then added. The hydrogen generated by the metal-acid reaction creates bubbles in the slip causing it to foam. As the solvent evaporates during the drying process, the pores formed by the hydrogen bubbles, which were originally closed, turn to interconnected porosity and an open cell green part is obtained. Figure 10 shows the typical structure for a steel foam which has irregular primary pore sizes as large as 3.5 mm and secondary pores between 0.05-0.3 mm. After sintering the total porosity is typically 60%. Foams have been produced from both steel and aluminium powders but for high porosities the green strength is low and cracks form in the foamed material.

In a variation on this process the slurry is an aqueous polymer solution that has the ability to form a gel (Shimizu and Matsuzaki, 2007). Gelation of the slurry is carried out by a freezing and thawing process and after gelation, it is heated until the foaming agent (hexane) decomposes, forming a gas, causing the gel to expand. The resulting foam is then dried and sintered.

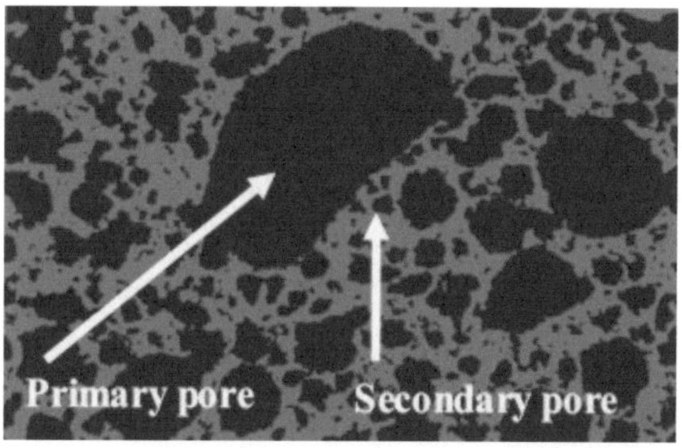

Fig. 10. Micrograph of a steel foam produced by the slip reaction foaming process (Angel et al., 2004).

2.2.4 Foaming by mechanical whisking

Foams have been produced from ceramic slurries by introducing air into the slurry, in much the same way as whisking cream (Sepulveda, 1997), and this method has recently been translated to metal systems (Lin, 2011). In the same way as for ceramics, in order to stabilise the air bubbles that are introduced during whisking, a surfactant must be added to the metal powder slurry. Through sufficient aeration, a foam can be formed which can be poured into a shaped mould made from almost any material. Despite the addition of surfactant, drainage of the liquid from the network of pores does occur, inevitably leading to collapse of the foam. To preserve the foam structure, slurry systems are designed to either gel by heating (cellulose systems) or cooling (agarose systems) or be polymerised by the addition of an initiator (acrylamide systems). The foamed body is then dried, further heated to burn out the polymer and finally sintered to densify the matrix.

The pore structures for these foams are surprisingly uniform given the simplicity of the process, showing round pores connected by small windows. Figure 11 shows a cross section through a stainless steel foam in the gelled and dried condition, which demonstrates good

Fig. 11. A cross section (left) of a gelled and dried foam before sintering and (right) the foam microstructure showing an open cell structure and (inset) the porous nature of the cell struts (Lin, 2011).

green strength, and also presents the sintered microstructures for the cells and the porous cell struts (inset). Porosities up to 90% have been achieved in sintered parts but work to date has shown that it is difficult to vary the pore size beyond the range of 0.5-1.5 mm. It is thought that this simple foaming process has the potential to produce foams that are suitable for both structural and functional applications.

2.3 Foaming of compacted powder precursors

2.3.1 Foaming process

Metal foams can be produced by encapsulating a foaming (or blowing) agent into a precursor made from compacted metal powder, followed by melting (Baumeister, 1990). The foaming agents are fine powdered compounds that, when heated, decompose to form a gas (typically they are metal hydrides or carbonates). When the compact is heated, usually in a mould, above the solidus temperature of the alloy, which should also be above the decomposition temperature of the foaming agent, the gas evolved causes expansion of the precursor. Expansion is rapid but collapse occurs, requiring fast cooling to "freeze-in" the foam structure. A schematic of the process used to make and foam powder precursors is shown in Figure 12.

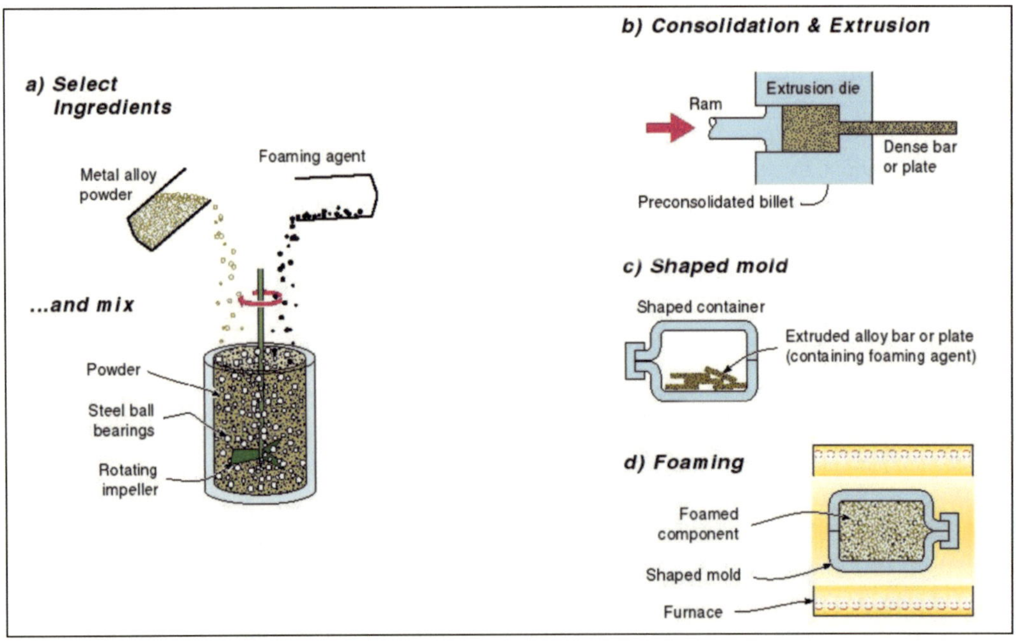

Fig. 12. The sequence of steps used to manufacture metal foams made from compacted powder precursors (Ashby et al., 2000).

Figure 13 shows an expanded precursor that has been foamed in a metal mould, thereby defining its cylindrical shape. Also shown is a radiograph of the sample, revealing the porosity inside the part. The cross section of the foam, also shown in this figure, reveals that although the pores are reasonably spherical, a large variation in the pore sizes is observed. The closed porosity in foams made in this way is typically below 90%.

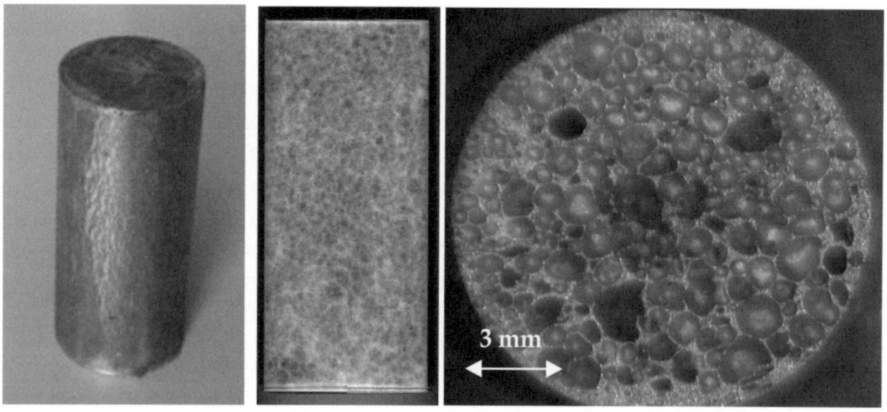

Fig. 13. Images of (left) a foam produced by the expansion of a melted PM precursor in a mould, (centre) an X-ray radiograph of the foam structure and (right) a cross section of the foam showing the pore structure.

2.3.2 Advantages and limitations of the process

The foaming of compacted powder precursors has several key advantages. It is one of the few processes that can make closed cell foams, with the ability to produce near-net-shape and complex foam parts, including foam plates and sandwich structures. Lightweight Al foam parts made in this way are being used in a number of applications where their high strength at low mass and excellent energy absorbing ability are exploited. The main disadvantages are that large 3D parts are difficult to make and, in part due to the rapid nature of the foam expansion process, the uniformity and reproducibility of the pore structure can be unsatisfactory, leading to concerns over the reproducibility of the mechanical performance (Kennedy, 2004).

Limits to the range of metals that can be foamed in this way arise due to several factors. There is a requirement for the blowing agent to only start to decompose when the metal or alloy is semi-solid. If decomposition occurs when the precursor is still solid, gas evolution can cause cracking of the precursor and escape of the gas without it contributing to foam expansion. It is, therefore, difficult to find suitable foaming agents for all metals. TiH_2, which starts to decompose at around 450°C (Kennedy and Lopez, 2003), is used to foam low melting point metals, such as Al and Mg, despite decomposing below the melting point of most of their alloys. This requires the compacted precursor to contain very low (<2%) porosity to contain the gas (Kennedy, 2002). This is not readily achieved by conventional PM compaction methods and so to achieve the target densities, the use of high strength, pre-alloyed powders is avoided, favouring mixed elemental powder additions, and powder consolidation is either performed by hot die compaction, cold isostatic pressing followed by extrusion or continuous powder rolling or extrusion processes. The foaming of reactive metals and those with high melting points is not that practical, given the need for a conductive metal mould to produce shaped parts, but $SrCO_3$ has been used to produce Fe-based foams.

Foams produced by melting compacted powders are stabilised (at least temporarily) by oxide films introduced into the liquid from the surface of the metal powders. The fraction of these oxides in the expanding liquid is critical to achieving good foam structures and stable foams. Variations in oxide content for different powder sizes, from different suppliers or even for different batches of powders (due to variations in the atomising or storage conditions) can be the reason behind highly variable foaming responses (Asavavisithchai and Kennedy, 2006a). Varying the processing conditions during hot compaction can also affect the oxidation of the metal powder and alter the foaming behaviour (Asavavisithchai and Kennedy, 2006a, 2006b). Figure 14 shows the effect of oxygen content on the foam expansion for an Al powder, too low and foam collapse is severe, too high and the liquid is too viscous to foam. It should be noted that the way in which this oxygen (or oxide) level is achieved, through heat treatments at different temperatures or through atomisation, is not important.

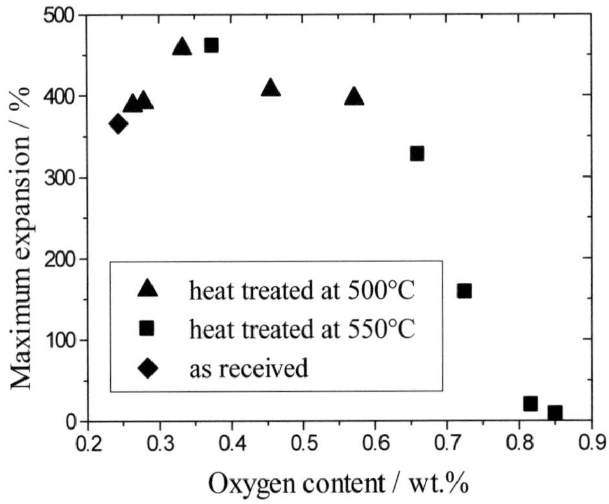

Fig. 14. The effect of oxygen content on the foaming of pure Al powder compacts, showing that an optimum level is required (Asavavisithchai and Kennedy, 2006a).

3. Summary

An overview of some of the many and varied methods for making porous metals and metal foams from metal powders has been presented. With research and development into metal foams being vibrant and dynamic, pioneered by institutes like the Fraunhofer IFAM centres in Bremen, Chemnitz and Dresden in Germany, the state-of-the-art is continually evolving as the understanding behind powder processing, foaming and foam stabilisation improves.

For established foaming processes, research conducted within academia and industry has a strong emphasis on eliminating problems which would otherwise limit the wider use of these materials. This includes; improving the uniformity and reproducibility of the foam structures, aiming to achieve uniform pore sizes and densities throughout the component and similar foam structures from part to part; decreasing the processing and materials costs, through improved or new processing routes, reduced waste and cheaper starting materials and developing compelling case studies based on innovative design, simulation and testing to demonstrate to end users that despite higher prices for some foams or components containing foam elements, that these costs can be more that offset by the weight and energy savings offered by these novel materials and structures.

4. References

Andersen O, Stephani G. 1999. Melt extracted fibres boost porous parts, *Metal Powder Report*, 54, 30-34.

Andersen O, Waag U, Schneider L, Stephani G, Kieback B. 2000. Novel metallic hollow sphere structures, *Advanced Engineering Materials*, 2,192.

Andersen O, Studnitzky T, Bauer J. 2004. Direct typing: a new method for the production of cellular P/M parts. In: Danninger H, Ratzi R, editors. *Euro PM2004 Conference Proceedings*. Shrewsbury: European Powder Metallurgy Association , vol 4. p.189

Angel S, Bleck W, Scholz PF, Fend T. 2004. Influence of powder morphology and chemical composition on metallic foams produced by Slip Reaction Foam Sintering (SRFS)-process. *Steel Res Intl*, 75, 483-488.

Asavavisithchai S., Kennedy A.R., 2006. Effect of powder oxide content on the expansion and stability of PM-route Al foams, *Journal of Colloid and Interface Science*, 297, 715-723.

Asavavisithchai S., Kennedy A.R., 2006. The Role of Oxidation During Compaction on the Expansion and Stability of Al Foams Made Via a PM Route, *Advanced Engineering Materials*, 8, 568-572.

Ashby MF, Evans A, Fleck NA, Gibson LJ, Hutchinson JW, Wadley HNG. 2000, *Metal Foams: A Design Guide*. Butterworth - Heinemann, UK.

Banhart, J., 2001. Manufacture, characterisation and application of cellular metals and metal foams. *Progress in Materials Science*,. 46(6), 559-632.

Baumeister J. 1990. German Patent 4,018,360,.

Eisenmann M. 1998. Metal powder technologies and applications. In: *ASM Handbook, vol. 7.* Materials Park, USA: ASM International,1031.

Elzey DM, Wadley HNG. 2001. The Limits of Solid State Foaming, *Acta Mater*, 49, 849.

Goodall, R., Mortensen, A., 2007. Microcellular aluminium: Child's play. *Advanced Engineering Materials*, 9, 951-954.

Jinnapat A, Kennedy A.R, 2010. The manufacture of spherical salt beads and their use as dissolvable templates for the production of cellular solids via a powder metallurgy route, *Journal of Alloys and Compounds*, 499, 43-47.

Kanetake N., Kobashi M., 2006. Innovative processing of porous and cellular materials by chemical reaction, Scripta Materialia 54, 521–525

Kennedy A.R., 2002. The effect of compaction density on the foamability of Al-TiH_2 powder compacts, *Powder Metallurgy*, 45, 1, 75-79

Kennedy A.R., Lopez V.H., 2003. The decomposition behavior of as-received and oxidized TiH_2 foaming-agent powder, *Mat Sci and Eng*, A357, 1-2, 258-263.

Kennedy A.R., 2004, Aspects of the reproducibilty of mechanical properties in Al based foams, *J Mat Sci*, 39, 3085-3088.

Kennedy, A R, Lin, X, 2011. Preparation and characterisation of metal powder slurries for use as precursors for metal foams made by gel casting *Powder Metallurgy*, 54,3, 376-381.

Lin X, 2011. Foaming of stainless steel powder slurries, *PhD thesis*, University of Nottingham, UK

Martin RL, Lederich RJ. 1992. Metal Powder Report, October, 30

Quadbeck P, Stephani G, Kuemmel K, Adler J, Standke G. 2007 Synthesis and properties of open-celled metal foams. *Mater Sci Forum*, (534-536): 1005-1008.

Sepulveda P. 1997. Gelcasting foams for porous ceramics. *Am Ceram Soc Bull*, 76, 61-65.

Shimizu T, Matsuzaki K. 2007. Metal foam production process using hydro-gel and its improvement. *Mater Sci Forum*, (539-543): 1845-1850.

Zhao, Y.Y., Sun D.X., 2001. A novel sintering-dissolution process for manufacturing Al foams. *Scripta Materialia*,. 44, 1, 105-110.

Zhou, J., 2006. *Advanced structural materials*. Porous Metallic Materials, ed. e. Winston O. Soboyejo., CRC Press , Taylor & Francis Group. USA, 22.

3

Aluminium Alloy Foams: Production and Properties

Isabel Duarte and Mónica Oliveira
Centro de Tecnologia Mecânica e Automação, Departamento de Engenharia Mecânica, Universidade de Aveiro, Portugal

1. Introduction

Ultra-light metal foams became an attractive research field both from the scientific and industrial applications view points. Closed-cell metal foams, in particular aluminium alloy (Al-alloy) ones can be used as lightweight, energy-absorption and damping structures in different industrial sectors, detaining an enormous potential when transportation is concerned. Despite the several manufacturing methods available, ultra-light metal foam applications seem to be restricted to a rather less demanding market in what concerns final product quality. The current available manufacturing processes enable effective control of density through process parameters manipulation. However, none of them allow for appropriate control of the cellular structures during its formation, leading to severe drawbacks in what concerns final product structural and mechanical properties. The latter, is undoubtedly the main reason for the lack of commercial acceptance of these ultra-light metal foams in product quality highly demanding sectors, such as the automotive or aeronautical sectors. The resolution of this problem is the main challenge of the scientific community in this field. To accomplish the latter two approaches may be followed: i) to develop new manufacturing processes or modify the existing ones to obtain foams with more uniform cellular structures. (ii) to understand and quantify the thermo-physico-chemical mechanisms involved during the foam formation in order to control the process, avoiding the occurrence of such imperfections and structural defects.

Metal foams manufacturing processes seem to abound (Banhart, 2001) and can be classified in two groups (Banhart, 2006): direct and indirect foaming methods. Direct foaming methods start from a molten metal containing uniformly dispersed ceramic particles to which gas bubbles are injected directly (Körner et al, 2005), or generated chemically by the decomposition of a blowing agent (e.g. titanium hydride, calcium), or by precipitation of gas dissolved in the melt by controlling temperature and pressure (Zeppelin, 2003). The indirect foaming methods require the preparation of foamable precursors that are subsequently foamed by heating. The foamable precursor consists of a dense compacted of powders where the blowing agent particles are uniformly distributed into the metallic matrix.

Most commercially available metal foams are based on alloys containing (Degischer & Kirst, 2002): aluminium, nickel, magnesium, lead, copper, titanium, steel and even gold. Among the metal foams, Al-alloys are commercially the most exploited ones due to their low density, high ductility, high thermal conductivity and competitive cost. Some manufacturing methods are already being commercialised. Direct foaming methods are currently being commercial exploited in a large-scale, the following companies are just a few examples. The Cymat Aluminium Corporation (in Canada) manufactures aluminium foams, designated "stabilised aluminium foam" which is obtained by gas injected directly into a molten metal (Degischer & Kirszt, 2002). Ceramic particles (e.g. silicon carbide, aluminium oxide and magnesium oxide) are used to enhance the viscosity of the melt and to adjust its foaming properties, since liquid metals cannot easily be foamed by the introduction of bubbling air. The foamy mass is relatively stable owing to the presence of ceramic particles in the melt. Foam panels with 1m in width and thickness range of 25-150 mm, can be produced continuously without length limitations at production rates of 900 kg/hour. The relative density of these foams is within the range 0.05-0.55 g/cm^3. The average cell size is 2.5-30 mm. This process is the cheapest of all and allows for manufacturing large volume of foams. Moreover, through this process it is possible to obtain low density foams. The main disadvantage lies is the poor quality of the foams produced. The cell size is large and often irregular, and the foams tend to have a marked density gradient. Despite this process continuous improvement, the drawing of the foam and the size distribution of the pores are still difficult to control. Besides its low production cost, secondary operations are usually necessary. For example, the foamed material is cut into the required shape after foaming. The machining of these foams can be problematic due the high content of ceramic particles (10-30 vol. %) used in the process. Hütte Klein-Reichenbach Ges.m.b.H company (Austria) also produces and commercialises aluminium foams with excellent cell size uniformity, called MetComb. The process used is based on the gas injection method (Banhart, 2006) and allows for the production of complex shaped parts by casting the formed foam into the moulds.

An alternative way for foaming melts directly is to add a blowing agent to the molten metal. The blowing agent decomposes under the influence of heat and releases gas which then propels the foaming process. Shinko Wire Company has been manufacturing foamed aluminium under the registered trade name "Alporas" with production volumes reported as up to 1000 kg of foam per day, using a batch casting process (Miyoshi et al, 2000). This method starts with the addition of 1.5 wt.% calcium metal into the molten aluminium at 680°C, followed by several minutes stirring to adjust viscosity. An increase of viscosity is achieved by the formation of calcium oxides. After the viscosity has reached the desired value, titanium hydride (TiH_2) is added (typically 1.6 wt.%), as a blowing agent by releasing hydrogen (H_2) gas in the hot viscous liquid. The melt starts to expand slowly and gradually fills the foaming vessel. The foaming takes place at constant pressure. After cooling the vessel below the melting point of the alloy, the liquid foam turns into a solid Al foam. After that, the foam block is removed from the mould, it is sliced into flat plates of various thicknesses according to its end use. This process is capable of producing large blocks of good quality. Blocks with 450 mm in width, 2050 mm in length and 650 mm in height can be produced. These foams have uniform pore structure and do not require the addition of ceramic particles, which makes it brittle. However, the method is more expensive than foaming melts by gas injection method requiring more complex processing equipment. The density range of these foams is 0.18-0.24 g/cm^3, and the mean cell size is about 4.5 mm.

Nowadays, foams manufactured by indirect foaming methods are also in the state of commercial exploitation, but in small-scale by German and Austrian Companies, like Schunk GmbH, Applied Light-weight Materials ALM and Austrian Company Alulight GmbH (Banhart, 2006). Powder Metallurgical (PM) method is one of the commercially exploited indirect methods to produce Al-alloy foams and it is also the research field of the authors of this chapter. This process consists on the heating of a precursor material which is obtained by hot compaction of a metal alloy (e.g. Al-alloy) with blowing agent powders (e.g. TiH_2), resulting in the foam itself. The metal expands, developing a highly internal porous structure of closed-cells due to the simultaneous occurrence of the melting of the metal and thermal decomposition of the blowing agent with the release of a gas (e.g. H_2). The liquid foam is then cooled in air, resulting in a solid foam with closed cells and with a very thin dense skin that improves the mechanical properties of these materials. This process can produce foams with porosities between 75% and 90%.

The PM method has several advantages in comparison with the methods described earlier and will be further discussed ahead. The latter, has been addressed particularly in what concerns two research lines: (i) the study of the physics and foaming technology, with particular emphasis on dedicated process equipment development towards high quality foams production. (ii) foam quality assessment through proper part property characterisation, establishing its limits of application and seeking for new markets. This chapter presents a detailed overview of the current state-of-art in what concerns to methods, equipment and appropriate industrial procedures in order to obtain metal foams with good quality. The advantages, the disadvantages and the limitations of this PM method are also presented and discussed. An overview of the main challenges and perspectives in this field concerning industrial implementation is also presented, whenever it is possible the authors present novel research work.

2. Production of metal foams

Metal foam production by PM method can be divided into two production steps: (i) production of foamable precursor and (ii) production of the metal foam itself through the foaming of precursor material. A schematic diagram of the PM method is shown in Fig. 1. The first step is the preparation of a dense solid semi finished product called foamable precursor. The latter is attained by compacting a powder mixture containing the blowing agent and the metal, by using a conventional technique. The second step includes the production of the metal foam by heating this foamable precursor at temperatures above its melting temperature.

This PM method can be used to produce foams of different metals and its alloys (Degischer &Kriszt, 2002), such as aluminium and its alloys, tin, zinc, lead, steel and gold, which is one of the advantage of this PM method. Among all metal foams, the Al-alloys are the ones that have received more attention from both the research community and industry, due to its enormous potential mainly in what concerns specific weight and highly corrosion resistance. The most studied Al-alloys for foaming are pure aluminium, wrought alloys (e.g. 6xxx alloy series) or casting alloys (e.g. AlSi7Mg). The high quality foams of these different metals can be obtained by choosing the appropriate blowing agent. Moreover, the manufacturing parameters of the different stages require appropriate adjustment (Fig. 2).

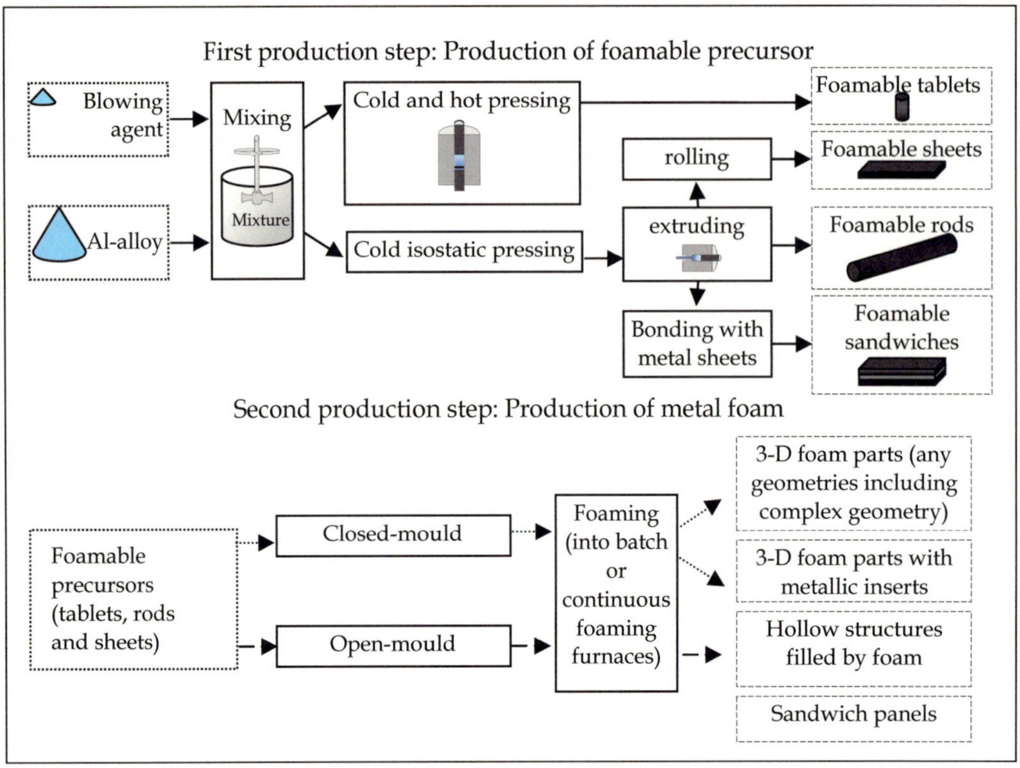

Fig. 1. Powder Metallurgical metod for making metal foams.

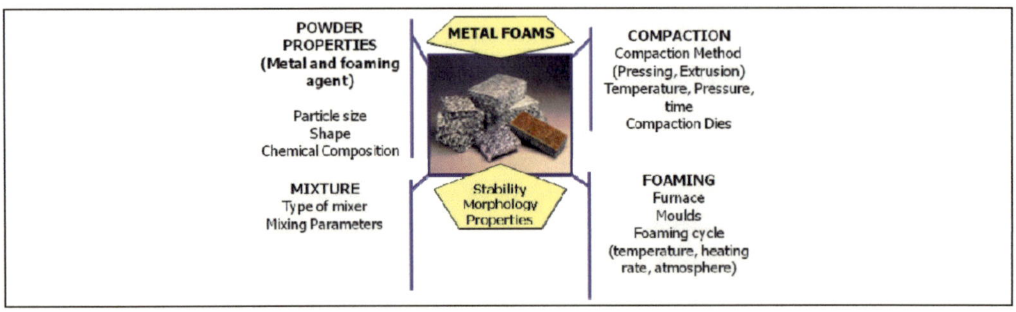

Fig. 2. Manufacturing parameters of PM method (Duarte, 2005).

The main advantage of this PM method is to enable the production of components of metal foams with different architectures (e.g. sandwich systems, filled profiles and 3D complex shaped structures) in comparison with the others (Duarte et al, 2008, 2010). The materials can be joined during the foaming step without using chemical adhesives (Duarte et al, 2006). Other advantage lies in the fact that the addition of ceramic particles are not required, avoiding the brittle mechanical behaviour inferred by these particles. Moreover, the foam parts are covered by an external dense metal skin that improves its mechanical behaviour, providing a good surface finish. The disadvantage of this PM process is the high production cost mainly associated to the powder prices. Another disadvantage is the difficulty to manufacture large volume foam parts. Nevertheless, sandwich panels of 2mx1mx1cm can already be manufactured (Degischer & Kirst, 2002). Furthermore, it must be pointed out that during PM method it is still rather difficult to fully control the foaming process, which results in lack of uniformity of the pore structure.

2.1 Preparation of the foamable precursors

The first step of the PM method to obtain metal foams is the production of foamable precursor materials. The fundamental aspects of this production step are discussed ahead.

2.1.1 Selection of the powders

The blowing agent is a chemical compound which releases gas when heated, being the responsible for the formation of bubbles. There are two main requirements to obtain high quality foams. The first one is to ensure an uniform distribution of the blowing particles into the metal matrix within the precursor material. The second is to ensure the coordination of the thermal decomposition characteristics of the blowing agent and the alloy melting behaviour to avoid cracks' formation before melting. The selection of these powders is therefore, detrimental for the foaming success. The characteristics of the metal powders, like the purity, the particle size, the alloying chemical elements (content and type) and the impurities (content and type), as well as the alloy melting behaviour and the thermal decomposition characteristics of the blowing agent must be studied and known. The literature highlights how powders from different manufactures could lead to notable differences in foaming behaviour (Baumgärtner et al, 2000; Degischer & Kirst, 2002).

The blowing agents usually used for producing Al-alloy foams using the PM method are metal hydrides, such as titanium hydride (TiH_2), zirconium hydride (ZrH_2) and magnesium hydride (MgH_2) (Duarte & Banhart, 2000). The employment of other blowing agent powders, such as the carbonates, has been investigated as a cost-effective alternative to metal hydrides (Haesche et al, 2010; Cambronero et al, 2009). It seems, though, that titanium hydride still is the best choice for the foaming agent when producing Al-alloy foams, as reported elsewhere (Duarte & Banhart, 2000). Fig. 3 presents a typical mixture of Al-alloy and TiH_2 powders. The amount of the metal hydrides usually used is less than 1% in weight in the initial powder mixture, based upon the Al or Al-alloy that is to be foamed. For example, high quality of AlSi7 foams can be obtained by using 0.6wt.% in the initial powder mixture (Duarte & Banhart, 2000).

The effect of the composition of the alloy, the impurities, the particle size and the alloying chemical elements on the foaming behaviour has been studied (Duarte & Banhart, 2000;

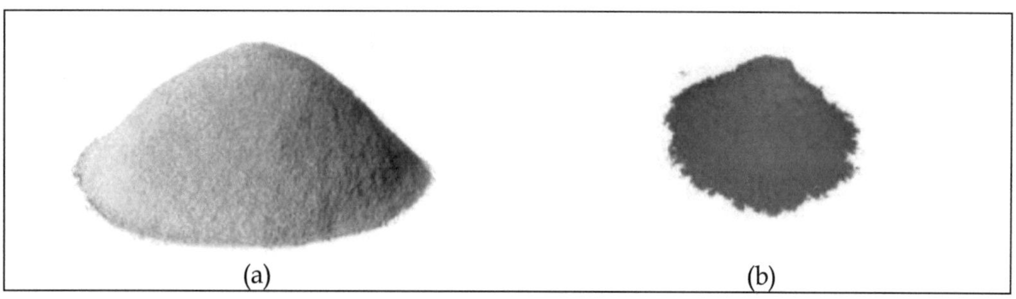

Fig. 3. Al-alloy and TiH_2 powders used in PM method.

Lehmus & Busse, 2004; Gokhale et al, 2007; Ibrahim et al, 2008; Helwig et al, 2011). Some of these effects on the foam expansion behaviour or cell structure have not yet been well established. A recent research of the effect of TiH_2 particle size on the foaming behaviour and on the morphology of Al-alloy foam produced by PM process is reported (Ibrahim et al, 2008). These studies revealed that the use of the coarser particle sizes of TiH_2 leads to a higher foam expansion and coarser macrostructure while the finer grade of TiH_2 leads to a quite lower maximum expansion and a finer macrostructure. The use of different particle sizes is an approach to adapt the onset of gas evolution temperature of the gas blowing agent and improvement of the macrostructure of foamed aluminium.

The difference between thermal decomposition of the blowing agent and melting temperature of the metal may cause the formation of irregular, crack-like pores in early expansion stages, which then can lead to irregularities in the final foam (Duarte & Banhart, 2000). The research in this field has been demonstrating that the high-quality Al-alloy foam is obtained when this difference is minimised. The basic rules to choose the best blowing agent are related to the closing between the temperature of the beginning of the thermal decomposition of the blowing agent and the *liquidus* temperature of the metal. This problem has been approached in two different ways: (i) pre-treatments of the blowing agent powder to delay the hydrogen release, i.e. to shift it to higher temperatures, (ii) to change the alloy composition to obtain lower melting point through the addition of alloying elements.

Thermal pre-treatments that lead to partial decomposition and/or pre-oxidation of the powder surface (Matijašević et al, 2006) or surface coatings (Proa-Flores & Drew, 2008) have been applied. For example, TiH_2 powder particles are heated (≈480ºC) in air during a given time (≈180 min) and an oxide layer is formed on the surface of the particles. This layer delays gas release from the particles, so that hydrogen is ideally released during foaming only after the alloy melting temperature has been reached. The powder colour depends on the thickness of the formed oxide layer (Fig. 4). Another example, the TiH_2 powders can be treated with acetic acid solution during 10 h, followed by a wash with distilled water until the solution presented a low acidity. The particles can then be coated with a silicon dioxide layer (Fang et al, 2005).

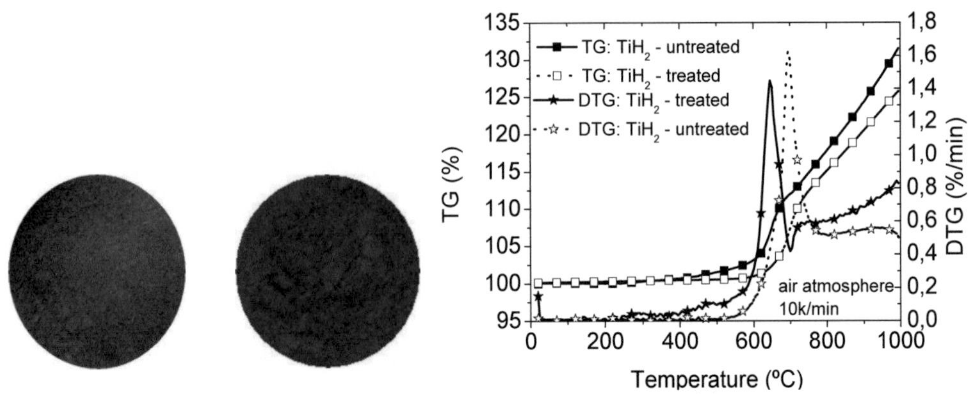

(a) (b) (c)

Fig. 4. Untreated (a) and treated (b) TiH_2 powders and their TG/DTG curves (c).

Another strategy is to change the alloy composition to obtain lower melting temperatures through the addition of alloying elements, such as the magnesium, zinc and copper that decreases the melting temperature (Helwig et al, 2011). Although this strategy appears to be promising, research in this field has not been very systematic. The results revealed that these treatments form a sufficiently thick oxide layer which leads to a minimum of hydrogen loose. Helwig *et al* reported that the use of even higher magnesium amounts was found out to lead to promising results in the run-up to this study. Researchers have been testing the addition of the ceramic particles (e.g. alumina) in the initial powder mixture to improve the foaming stability through the increase of the melt bulk viscosity (Kennedy, 2004). However, the presence of these particles can originate a brittle mechanical behaviour of the foams.

2.1.2 Mixing of the powders

The mixing procedure should yield a homogeneous distribution of the alloying elements and the blowing agent particles to ensure the high-quality of the Al-alloy foams with an uniform pore size distribution. Blending the Al-alloy and the blowing agent powders is a crucial step within the entire foaming process. This operation should be made to avoid the agglomeration of the blowing agent particles and the alloying elements. This causes structural defects and imperfections on the final foam. The mixers usually used are tumbling mixers with or without alumina balls (Fig. 5). These balls do not add any other element to the mixture, because the aluminium oxide is already present in the mixture.

An important practical aspect in the mixing operation is to obtain a clean and homogeneous powder mixture. The impurities and solid powders by dirt, water or other particles entrapped in the mixture may have a detrimental effect in foaming. These impurities can act like nuclei uncontrolled voids during the thermal decomposition of the blowing agent which will form larger pores at the latest foaming stages (Matijasevic & Banhart, 2006).

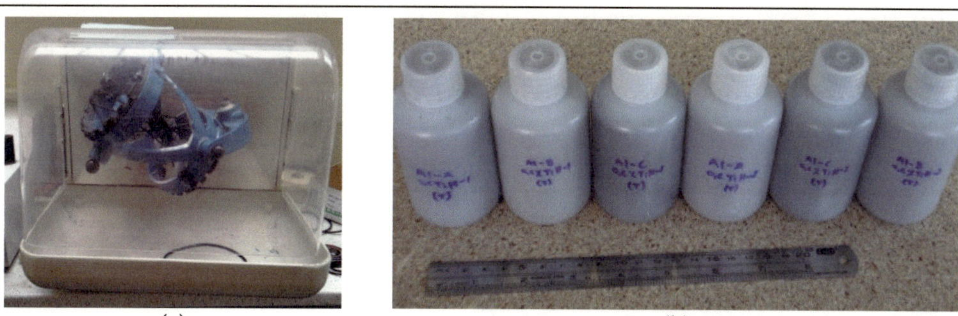

(a) (b)

Fig. 5. (a) Turbula mixer used to mix the powders. (b) Powder mixtures.

2.1.3 Compaction of the powders

The compaction should ensure that the blowing agent particles are embedded in the Al-alloy matrix. The basic practical rule is to obtain a dense semi compact called foamable precursor material with no residual open porosity in which theoretical density is close to 100% of the theoretical density of the aluminium matrix (Duarte&Banhart, 2000). The production of precursors by compacting powder mixtures can be performed in a variety of ways, e.g., by uni-axial (Kennedy, 2004), double-axial or isostatic pressing (Körner et al, 2000), extrusion (Baumgartner et al, 2000), powder rolling (Kitazono, 2004), etc., and all the above mentioned techniques can be hot or cold. A conjugation of different conventional compaction techniques can be used. Furthermore, the compacting process can be performed in an inert atmosphere, in air or in vacuum (Jiménez et al, 2009). The most economical way is the double-axial pressing in air, but the most efficient one is the high-temperature extrusion. There are several procedures to produce foamable precursors. A simple process is to compact the powder mixture using a hot uniaxial pressing at temperatures close to the thermal decomposition of the blowing agent, using a pressing device and a die with a heating system (Fig. 6). This process enables to yield more than 99% relative density of the precursor (Fig. 7a).

Fig. 6. Die with the heating system used to prepare the precursor material.

The initial powder mixture is first compacted to cylindrical billets of 70-80% of the theoretical density, using the cold isostatic pressing. Then, these billets are pre-heated to temperatures close to the thermal decomposition of the blowing agent and extruded into rectangular bars of various dimensions (Baumgärtner et al, 2000), as shown in Fig. 7b.

Kennedy reported that a minimum compaction density is required to achieve appreciable expansion and is about 94% of the theoretical density (Kennedy, 2002). The highest value of the expansion is obtained when the density is close to the theoretical density of the Al-alloy matrix. The authors research results reveal that compaction of the mixture to achieve high quality precursor materials must be performed at temperatures close to the initial thermal decomposition temperature (e.g. 400°C), in order to achieve the density value which leads to a good foaming behaviour Moreover, the cold compaction should be done before the hot-compaction in order to obtain a high level of densification (Fig.8).

(a) (b)

Fig. 7. Precursor materials of Al-alloy containing 0.6%wt. of TiH2 which is manufactured using the laboratory system in Fig.6 (a) and supplied by IFAM (Baumgärtner et al, 2000).

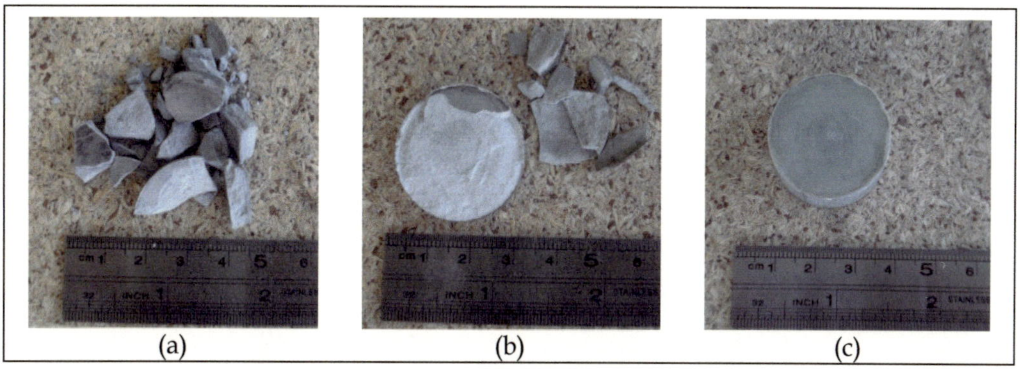

Fig. 8. Precursor materials at different densification levels: (a,b) low-quality (c) high-quality.

Several parameters are evaluated to ensure the quality of the foamable precursors. The main parameters are the density and the distribution of the blowing agent particles into the metal matrix. The latter is usually evaluated by scanning electron microscope (Fig. 9).

Fig. 9. Foamable precursor sample of aluminium alloys containing 0.6 wt.% of TiH_2 composed with Al-matrix (dark gray) with Si (light gray) and TiH_2 (white colour) particles.

The required density of the foamable precursor is adjusted by manipulating the compaction parameters (time and temperature). Duarte and Banhart reported that the hot-pressing temperature is a very important parameter (Duarte & Banhart, 2000). The reduction of compaction temperatures lead to insufficient compaction with some residual porosity (see density values given in Fig.10 a). The hydrogen gas can escape from the melting alloy without creating pores in this case. The same phenomenon is observed when the powders are extruded instead of hot pressed. On the other hand when higher compaction temperatures are employed lower maximum expansions can be observed, mainly because some of the hydrogen is lost during compaction. Even higher compaction temperatures lead to a rapid loss of foamability. The optimum compaction temperature therefore lies around 450°C for the case investigated, well above the initial decomposition temperature of TiH_2 (380°C). The compaction time is not considered a critical parameter for the compaction temperatures chosen. The variations observed are not systematic and seem to be within the range of normal statistical fluctuations.

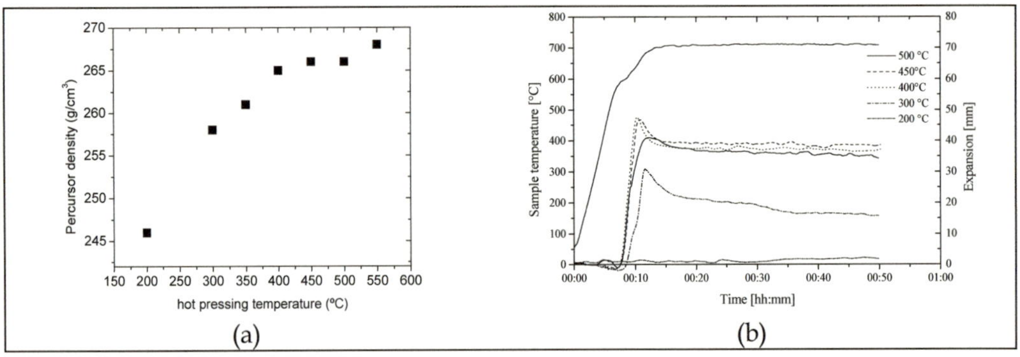

Fig. 10. Effect of the hot pressing temperature on the precursor density (a) and on the foaming behaviour (b).

2.2 Production of metal foams

The second step is the production of metal foams by heating the foamable precursor at temperatures above the melting point of the alloy. The metal expands developing a highly porous closed-cell internal structure due to the simultaneous melting of the aluminium and thermal decomposition of the blowing agent (TiH_2) in gas (H_2). The solid foams are obtained by controlled cooling of the formed liquid foam, at temperatures below the *solidus* temperature. The foamed parts have a highly internal structure with closed cells (Fig. 11 b) and are covered by a dense metal skin that improves their mechanical properties and provide good surface finish (Fig. 11).

The foaming process usually takes place into the stainless steel closed moulds (Fig. 12). The cavity of the mould should have the same design and dimensions of the final foam. The foamable precursor into the mould is heated at temperatures above its melting point (Fig. 12 b). The material expands and fills the entire mould cavity. After that, the mould is cooled, followed by the extraction of the foam (Fig. 12c). The furnaces used to manufacture the metal foams usually are of the batch chamber furnace type (Fig. 13a). This research team has developed a foaming continuous furnace to produce these materials (Fig. 13b).

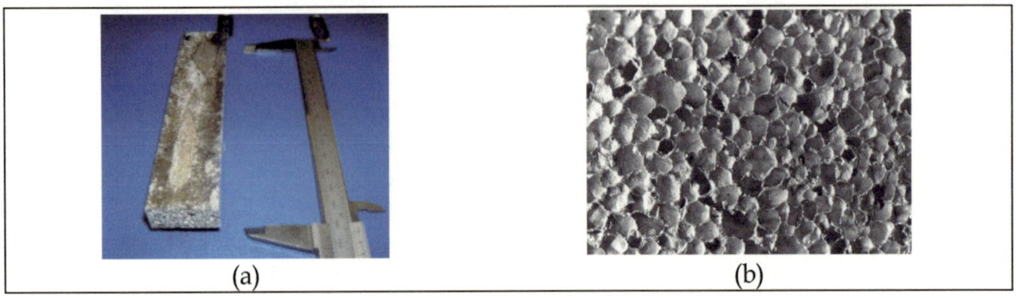

Fig. 11. Al-alloy foam covered by a dense skin (a), with a closed-cell internal structure (b).

This method enables the cost effective production of aluminium alloy parts without limitations concerning shape (e.g. panels, profiles or complex 3D shaped parts) (Duarte et al, 2006, 2008, 2010). Here, the group research results are presented to illustrate some practical examples (Fig. 14) which can be obtained by using the continuous foaming furnace developed (Fig. 13 b). The disadvantage of the process is to produce components with a large volume parts as it is the case of Shinko Wire method (Miyoshi et al, 2000).

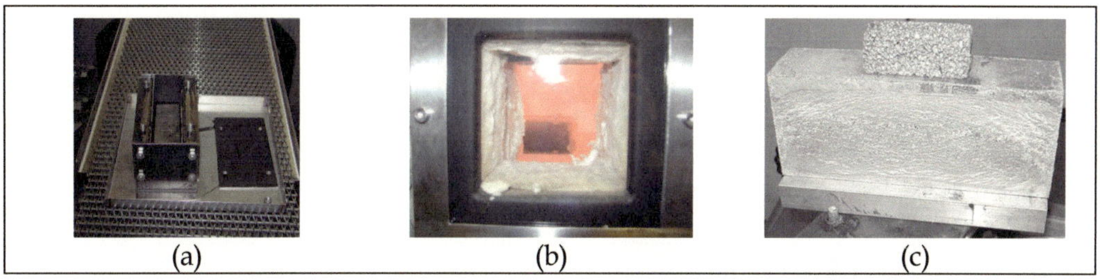

Fig. 12. Practical aspects of the process. (a) closed mould. (b) mould with the precursor into the pre-heated furnace. (c) Al-alloy foam block.

Fig. 13. Foaming furnaces. Batch chamber furnace (a). Continuous foaming furnace developed by this research team (b).

Fig. 14. 3D-parts of Al-alloy foams.

The atmosphere, heating rate and temperature of the thermal foaming cycle are some of the manufacturing parameters which influence the quality and the properties of the resulting foam. Moreover, it should also be referred that the characteristics of the mould (material, design and dimensions), the type of the furnace (batch or continuous) should also be considered (Duarte, 2005). The effects of these variables on the foaming behaviour may therefore be investigated. To assess the latter, foaming tests are performed by heating the precursor material up to its melting point inside of the apparatus called laser expandometer, which was specially developed and constructed for this purpose (Fig. 15). Here, the expansion (volume) and its temperature are monitored by means of a laser sensor and thermocouple, respectively. The measurement of the volume of the expanding melt together with the sample temperature generates a pair of functions V(t) and T(t) which characterise the foaming kinetics. The foaming process is strongly governed by temperature effects (Duarte & Banhart, 2000). The expansion curve is strongly dependent on the processing conditions, mainly on the hot pressing temperature to obtain the foamable precursor material, and the heating parameters during foaming (Fig. 10 b and Fig. 16).

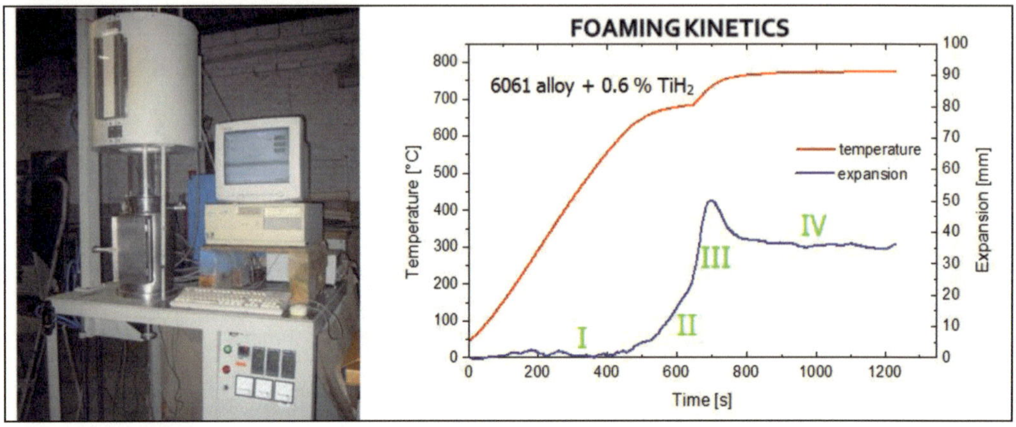

Fig. 15. Laser expandometer used to characterise the foaming kinetics.

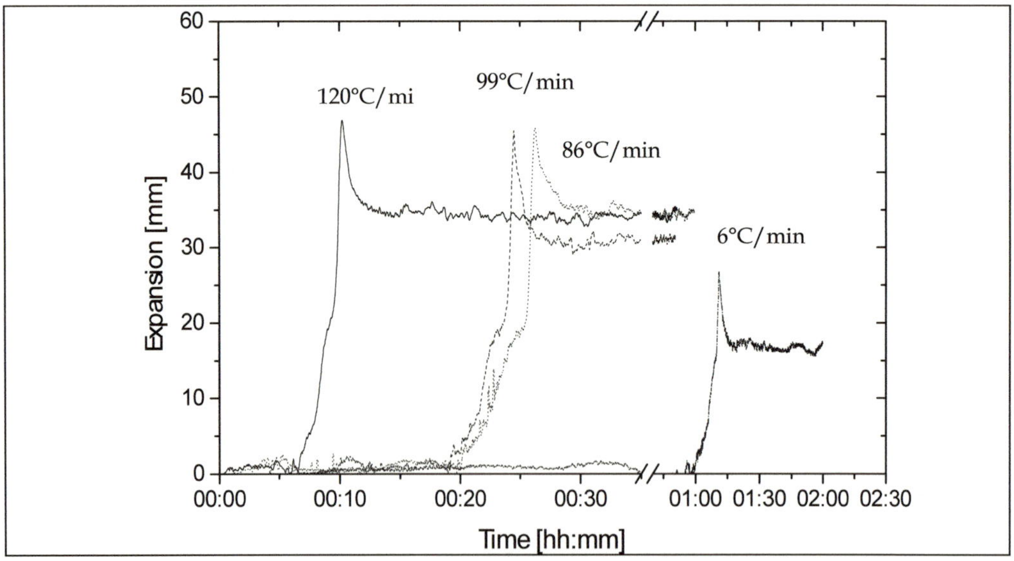

Fig. 16. Expansion curves of 6061 samples containing 0.6 wt. % TiH_2, prepared with different heating rates (Duarte & Banhart, 2000).

The final quality of the metal foams is evaluated by characterising its properties, namely the density, structural and mechanical properties. The foam density is relatively predictable and controlled by manipulation of the manufacturing parameters (Duarte & Banhart, 2005). Higher heating rates lead to an earlier expansion of the foamable precursor because the melting temperature is reached at an earlier time (Fig. 16). Besides that, the three expansion curves for the highest heating rates are quite similar. Only significantly lower heating rates lead to a change of the expansion characteristics, namely a lower maximum expansion. The reason for this may be the gas losses due to diffusion of hydrogen and perhaps the strong sample oxidation which might hinder expansion. In general, higher heating rate of the foamable precursor leads to the formation of foams with lower density.

Although at an industrial stage, the process still has major limitations on the ability to obtain tailor made cellular structure foams and to predict their properties. Cellular structured foams with different sizes and shapes of pores, and structural defects, can present a high density gradient depending on the component size (Fig. 17). These imperfections arise from the difficulties in controlling the manufacturing process such as: (i) lack of homogeneity in the precursor mixture (metal+blowing agent). (ii) lack of coordination between the mechanisms of metal melting and thermal decomposition of the blowing agent. (iii) difficulty in controlling the nucleation, growth and collapse stages. (iv) difficulty in stabilizing the foam formed and preventing the cells' collapse.

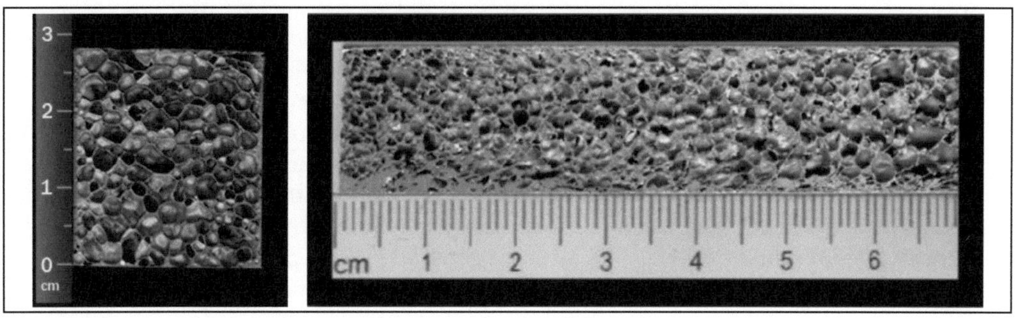

Fig. 17. Internal cellular structures for closed-cell AlSi7-alloy foam with different size.

3. The physics of metal foaming

The study of the physics of metal foaming is necessary to understand the underlying principle of metal foam formation and stabilization in order to produce better foams. This knowledge should lead to a control of these mechanisms during the foam formation. Currently, there are no detailed studies on the quantification of these mechanisms, only some aspects of foam evolution could be described by theoretical approaches. Despite the longstanding interest and research efforts regarding these mechanisms, the simulation and prediction of bubble nucleation in metallic foaming remains challenging. This is due to the difficulties in observing these mechanisms under experimental or actual processing conditions. For that, the researchers have been trying to quantify the involved mechanisms. Several investigations on foam formation have been carried out. The experimental techniques used for investigating the metallic foaming are mainly of two types: *ex-situ* techniques (Babcsán et al, 2005) and *in-situ* techniques (Babcsán et al, 2007). In the *ex-situ* techniques, the foaming process is interrupted by cooling at different foaming stages and the resulting solid foam is characterized. The disadvantage of this approach is that it takes a long time to carry out such investigation, and that the results suffer from a certain inaccuracy originating statistical variations between a single experiment. Even if the starting materials for the individual foaming tests were produced in the same way, each foaming experimental test would turn out slightly differently, due to effects such as agglomerates of the blowing agent particles, structural defects and impurities in the precursor (Duarte, 2005). The 3D-image of X-ray tomographic observations of solid foam samples in different foaming stages allow to observe the modification of the shape of the bubbles (Stanzick et al, 2002).

In the *in-situ* techniques, the foaming studies are evaluated during the evolution of one single sample (Garcia-Moreno et al, 2005; Rack et al, 2009). Neutron radioscopy and X-ray radioscopy were employed for *in situ* observation drainage mechanisms, the early stages and growth stages, during the foaming process (Bellman et al, 2002). The temporal development of the cellular structures of liquid metallic foams and the redistributions of the metal can be observed by synchrotron based on neutron radioscopy.

Theoretical studies of the foaming process itself concentrate predominantly on the detailed analysis of microscopic evolution of foams, mostly on the basis of an already existing cellular structure (Stavans, 1993). A theoretical study on metal foam processing has treated the material flow behaviour and focused on the description of the solidification stage during the process using an one-dimensional model which combines the equations of foam drainage with Fourier equation. However, this model is not able to describe the entire foaming process starting from bubble nucleation to final foam development. Other authors showed that the standard foam drainage equation (FDE) can principally be used to describe drainage in metal foams (Körner et al, 2008; Brunke & Odenbach, 2006; Belkessam & Fristching, 2003). However, the effective viscosity was considered to be one order of magnitude higher than the original one to match the experimental observations (Stanzick et al, 2002). A published model of metal foaming based on a Lattice–Boltzmann procedure (Körner et al, 2002) treats the foaming problem in more detail, (i.e. from bubble nucleation to the resulting foam structure, however, without considering the chemical decomposition of the blowing agent as well as the thermal heating process). The foam is considered in the liquid state; i.e, melting and solidification are not taken into account.

These studies have had significant roles in contributing to a more complete understanding of bubble-growth phenomena. However, in studying bubble-growth behaviour, almost all of these previous works involve pure theoretical studies without experimental verification, since only limited experiments have addressed the dynamic behaviour of the phenomena. Moreover, some of the physical parameters that were used to describe the materials adopted in these theoretical studies were unrealistic. There are no mathematical models available for foam evolution including nucleation, growth, coalescence and decay. Generally, numerical or analytical models focus on a particular phenomenon (e.g. drainage). The analytical and numerical approaches available in the literature are very limited. From the engineering point of view it is a very complex process because it involves several physical, chemical, thermal and mechanical phenomena that occur at the same time or successively.

3.1 Foam evolution

The evolution of the cellular pores during the foaming process has been evaluated and discussed using the *ex-situ* and *in-situ* techniques, as shown in the presented examples in Fig. 18 and 19, respectively.

The foaming process can be divided into three stages: bubble nucleation, bubble growth and foam collapse (Fig. 18b). Duarte and Banhart have done some pioneer research, in the understanding of the mechanisms involved in the PM process, which contributed to the significant advance of the state of the art (Duarte & Banhart, 2000), using *ex-situ* technique in which the foaming process is interrupted by cooling at different foaming stages and the resulting solid foam is characterized. Topics such as bubble nucleation, bubble growth and foam stability were discussed. As a result of this research, it is concluded that the bubbles nucleation occurs usually in the solid state, the pressure generated by the gas can deform the metallic matrix, the process is governed by the principle of semisolid metal processing, the bubble growth dynamics is controlled by the thermal decomposition of the blowing agent and the metal melting.

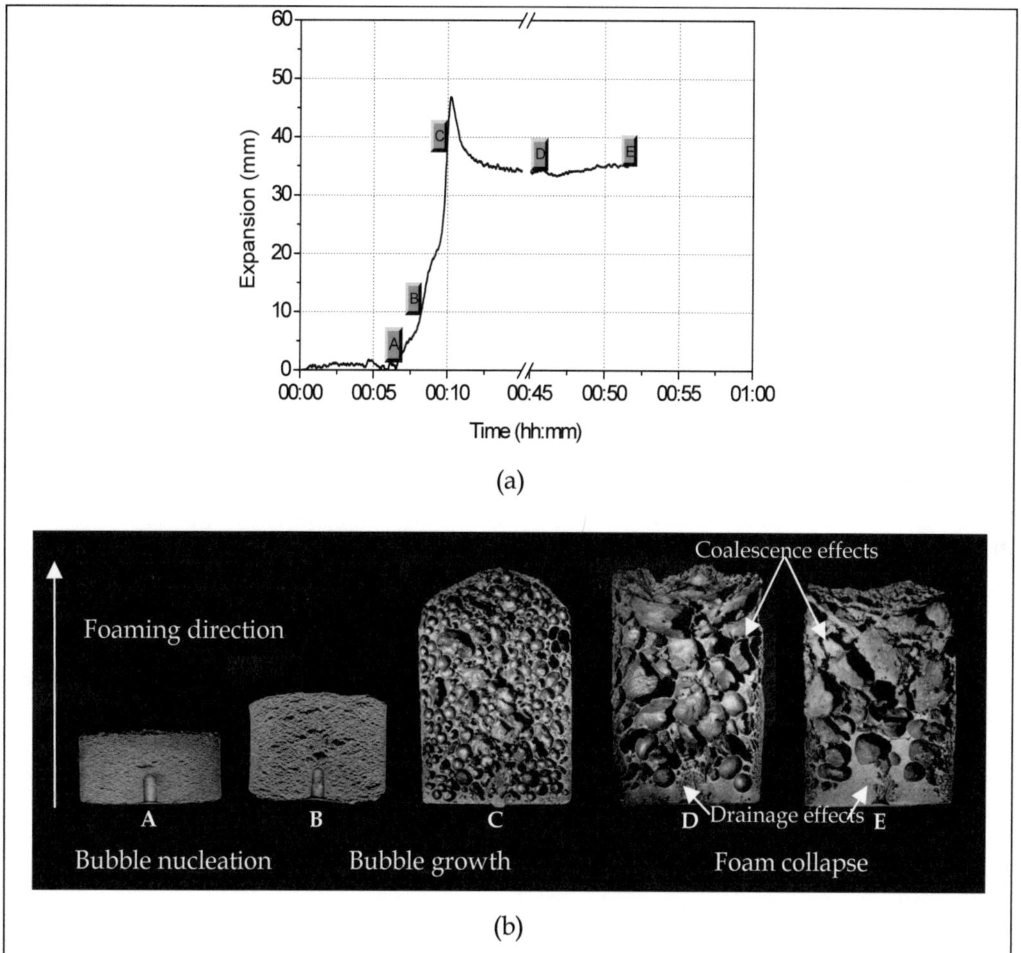

Fig. 18. (a) Expansion curve of precursor containing AA 6061 sample containing 0.6 wt. % TiH_2 foamed in a pre-heated furnace at 800°C. (b) Morphology of the AA 6061 foams in different foaming stages (foam diameter ≈ 30 mm). The letters show at which expansion stage the sample was removed (see Fig. 15).

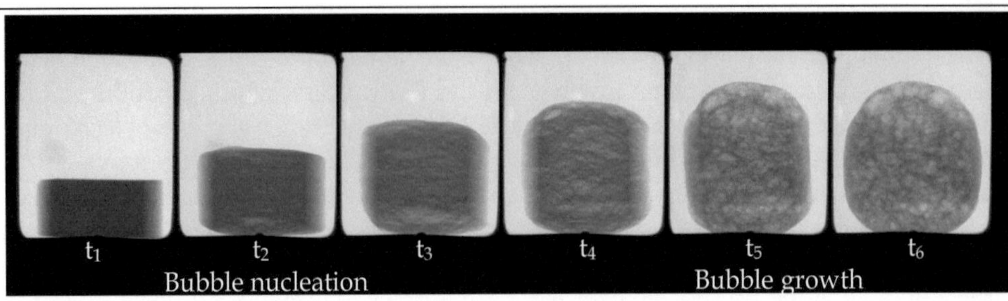

Fig. 19. In-situ radioscopic images acquired during a foaming process of a precursor.

The foaming process can be divided into three stages: bubble nucleation, bubble growth and foam collapse (Fig. 18). The shape of the bubbles varies during the foaming process. The bubbles appear as cracks aligned perpendicular to the foaming direction, changing to a spherical geometry, followed by polyhedral geometries. As described above, the PM method consists on heating a solid precursor material (Fig. 9). When heated, the metal melts (e.g. Al alloy) and the blowing agent (e.g. TiH_2) produces a gas that creates the bubbles in the foam. The heating process leads to partial metal melting as well as to the release of the gas and consequently to the material foaming in its semi-solid state of the material. The heat supply takes place in the solid material up to the *solidus* temperature. The decomposition of the blowing agent may start in the still with the precursor in solid state, so that the bubble nucleation can be previously initiated here. For this reason, the pores formed at early stages of foaming appear as cracks aligned perpendicular to the foaming direction (Fig. 18b in the nucleation stage). Moreover, the partial decomposition of the blowing agent particles during the compaction step can occur in which the temperature in this step is near to the initial decomposition temperature. The released gas in this compaction step is entrapping into the matrix metal, and can create a sufficient number of initial nuclei into the foamable precursor in which can act as centre of bubble nucleation. Gas accumulates in residual porosity and builds up pressure as temperature increases. With the occurrence of bubble growth, a two-phase flow (liquid metal and gas bubbles) develops above the *solidus* temperature.

After the alloy melting, the crack-like pores round off to minimize surface energy. Bubble growth begins, driven by gas release from the blowing agent, and the structure starts to appear as foam. With the increase of the temperature the internal gas pressure of each nucleated bubble increase, and turn in the strength of the metal matrix is reduced down to its value at the melting point. The bubble growth may not be uniform because depends on the characteristics of the foamable precursor material. The elongated initial bubbles are increasing in size and become more spherical (Fig. 18). The more spherical bubbles are observed in the cellular structures when the expansion reaches the maximum value. The spherical bubbles become more polyhedrical bubbles. After maximum expansion, no more gas is released and the foam begins to collapse. The latter is due the drainage and coalescence mechanisms, which are discussed ahead.

Foam growth depends not only on the rheology of the system metal/gas and mechanical strength but also on the pressure inside each bubble and its architecture. This growth is affected by various factors such as the content and distribution of the blowing agent, the hydrostatic pressure or tension applied to the metallic matrix and the viscous properties of the system (metal+gas). The metallic foaming rheology is not simple and its mechanical, thermal and chemical interaction leads to coupled problems of great complexity.

3.2 Foam: Collapse, stabilisation and solidification

The fundamental stability, collapse and solidification mechanisms during the foam formation have been investigated. These topics are the most controversial ones in the metal foam research. A better understanding of these mechanisms is required for an accurate control of the foam structures, such as cell size and porosity. *Ex-situ* and *in-situ* techniques have been used in this field. *In-situ* techniques have been used to measure the expansion, the density evolution and the effects of the drainage and coalescence. The effects of the thermal foaming cycle (heating rate, temperature and atmosphere) on the foam stability and foam collapse parameters have been also studied and observed (Banhart, 2006). PM foams belong to the class of transient (unstable) foams with lifetimes of seconds or to the permanent (metastable) foams with lifetimes of hours. The foamability is thought to result from the Gibbs-Marangoni effect where a membrane is stabilized during thinning due to liquid metal flow towards the weakened region, because of a local increase in surface tension. The flow is the response to a surface tension gradient. Due to viscous drag the flow can carry an appreciable amount of underlying liquid along with it so that it restores the thickness. In contrast to transient foams, the film thinning times in permanent foams are relatively short compared to the lifetime. The stability is controlled by the balance of interfacial forces. These forces equilibrate after drainage has been completed. The effect of temperature and gravity have been evaluated via observation of bubble size, ruptures of the bubbles, relative distribution of ruptures between bottom and top, draining and timing of draining, as well as the effect of temperature distribution.

Foam decays by combination of three phenomena - gas loss, drainage and coalescence (Duarte, 2005). Part of the gas is lost by diffusion from the outer surface to the surrounding. In the liquid state, loss is expected to rise due to a higher diffusivity than in its solid state. However, crack formation during expansion at the outer surface of precursors can also lead to gas losses. Moreover, bubble rupture at outer surface also results in sudden gas loss. The second and third way of gas loss are random events and not much effort was attributed to asses their contribution except for some qualitative statement. Drainage is one of the driving forces for the temporal instability of liquid foams, caused mainly by gravitational and capillary effects. The drainage and coalescence mechanisms can be observed using *ex-situ* techniques with the formation of a thick metal layer on the bottom of the solid foam and large pores, respectively. Foam shows a very complex rheological behaviour including the bubble deformation, rearrangement and avalanches processes.

Foam solidification is also an essential processing step of foam production. An uncontrolled solidification can create defects (e.g. cracks) in the cell wall of the foams (Duarte, 2005). So, this study helps to reduce the defects observed in final structures and thereby to improve the mechanical performance of the foam. The solidification by cooling of a foamed liquid metal is a "race against time", in as much as the relatively heavy liquid is prone to drainage, which rapidly reduces the foam density and hence provokes instability and collapse. The foamed liquid is immediately subjected to gravity-driven drainage of liquid, creating a vertical profile of density (or liquid fraction). At any point in the sample this adjustment must proceed until the freezing point is reached. Thus, at intermediate times, the sample consists of a solidified outer shell surrounding a draining liquid core. The competition between drainage and heat transfer, leading to solidification was studied by Mukherjee *et al* using X-ray radioscopy (Mukherjee et al, 2009). A hitherto unknown expansion stage was observed during solidification of Al-alloy foams. The phenomena that occur simultaneously while foam solidifies are associated to its volume change (Mukherjee et al, 2009). The extra expansion is observed and it is an anomalous behaviour since the foam is expected to contract during solidification. This extra expansion takes place whenever the combined volume gain rate is more than the combined volume loss rate. Moreover, it increases as the cooling rate decreases. Mukherjee reported that the slow cooling of foams can trigger extra expansion which in turn can induce defects in foam morphology. This is due to the extra expansion induced rupture during solidification of the metallic melt inside the cell wall.

4. Properties

The properties of the metal foams belong to a group of materials called cellular solids which are defined as having porosity >up to 0.7 (Gibson & Ashby, 1997). Natural foams are produced by plants and animals such as cork or bone. Man made foams can be manufactured from a variety of materials such as ceramics, polymers and metals. There are two categories of foams: open - and closed- cells. Here, it is presented a brief overview of the main properties of closed-cell Al-alloy foams obtained by the PM method.

Metal foams combine properties of cellular materials with those of metals. For this reason, metal foams are advantageous for lightweight constructions due to their high strength-to-weight ratio, in combination with structural and functional properties like crash energy absorption, sound and heat management (Asbhy et al, 2000; Degischer & Kriszt, 2002). Many metals and their alloys can be foamed. Among the metal foams, the Al-alloy ones are commercially the most exploited due to their low density, high ductility, high thermal conductivity, and metal competitive cost.

4.1 Structural properties

There are several structural parameters of these foams, such as number, size-pore distribution, average size, shape and geometry of the pores, thickness, intersections and defects in the cell-walls and thickness, defects and cracks of the external dense surface for describing the cellular architecture of the foams. The properties of these foams are influenced by these morphological features (Gibson &Ashby, 1997; Ramamurty & Paul, 2004; Campana et al, 2008).

Progress has been made in understanding the relationship between properties and morphology. Although this exact interrelationship is not yet sufficiently known, one usually assumes that the properties are improved when all the individual cells of a foam have similar size and a spherical shape. This has not really been verified experimentally. There is no doubt that the density of a metal foam and the matrix alloy properties influence the modulus and strength of the foam. All studies indicate that the real properties are inferior than the theoretically expected due to structural defects. This demands a better pore control and reduction in structural defects. Density variation and imperfections yield a large scatter of measured properties, which is detrimental for the metal foams reliability (Ramamurty & Paul, 2004). Wiggled or missing cell-walls reduce strength, and in turn, result in a reduced deformation energy absorbed under compression (Markaki & Clyne, 2001; Campana & Pilone, 2008). Mechanical studies demonstrate that selective deformation of the weakest region of the foam structure leads to crush-band formation (Duarte et al, 2009). Cell morphology and interconnection could also affect thermal and acoustic properties (Kolluri et al, 2008). It is widely accepted that foams with a uniform pore distribution and defects free, are desirable. This would make the properties more predictable. Only then, metal foams will be considered reliable materials for engineering purposes and will be able to compete with classical materials. Despite their quality improvement in the last 10 years the resulting metal foams still suffer from non-uniformities. Scientists aim to produce more regular structures with fewer defects in a more reproducible way which is the crucial challenge of the research in this field.

Foam characterisation results revealed that the cellular structures of the Al-alloy foams obtained by PM method have pores with different sizes and shapes (Fig. 20). A large size distribution of the cellular pores with irregular cell shape is observed. The closed pores are- mostly- of polyhedral or spherical geometry (Fig. 20 and 21). Spherical pores with a thick thickness of cell-wall are mostly observed in the bottom and lateral sides of the foam samples (Figs. 20 and 21a). Polyhedral pores with a thin cell-wall thickness are mainly observed at the top of the foam samples (Figs. 20 and 21b). The distribution of the solid metal in the foam is also non-uniform and leads to a higher density gradient (Fig.20). These materials have a broader cell diameter distribution curve (Fig. 20c). The cell-size distribution is dominated by high number of the small pores. The most of the pores have diameter lower than 2mm. The magnified images of the cross section of the sample reveal small porosities in the dense surface skin (Fig. 22). Significant morphological defects such as cracks or spherical micropores in cell walls and cell wall wiggles and dense surface skin are also observed. Each cell has normally approximately 5 other ones in its vicinity (Fig. 21). The distribution of the cell-wall thickness has an asymmetric shape for these foams (Fig.21). The smallest cell-wall thickness is about 70 μm. The maximum cell-wall thickness is about 500 μm. The thickness of the cell wall depends on the foam density. The thickness of the external dense surface skin around sample varies, where the higher values are located in the lateral and bottom sides. AlSi7 foams presents 565.56 μm, 365.40 μm and 214.58 μm, respectively for bottom, lateral and top sides of the samples (Fig. 22). Other structural feature that affects the mechanical behaviour is the microstructure of the massive cell material. Depending on the

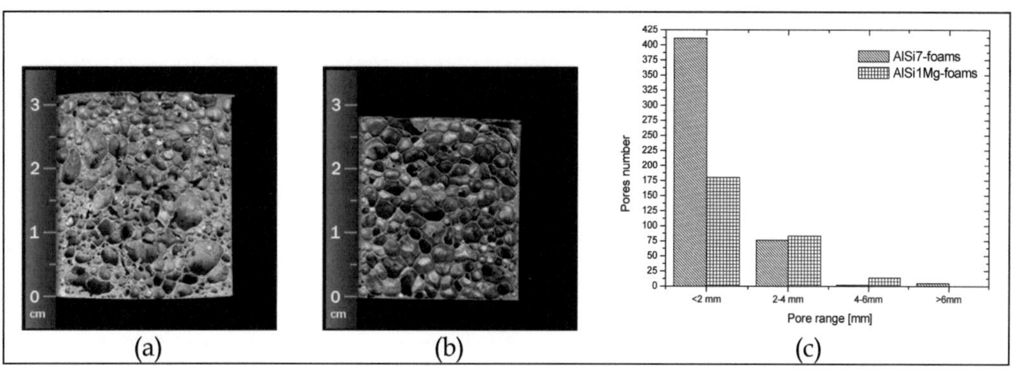

Fig. 20. Cellular structures of AlSi7 (a) and AlSi1Mg (b) foams. Cell-pore size distribution as a function of the number of cell pores for both foams (c).

alloy composition and on the manufacturing process, metallic dendrites, eutectic cells, precipitates, or even particles can be observed in the final cellular structures. Foams having same density but made of different Al-alloys can reached different plateau stress.

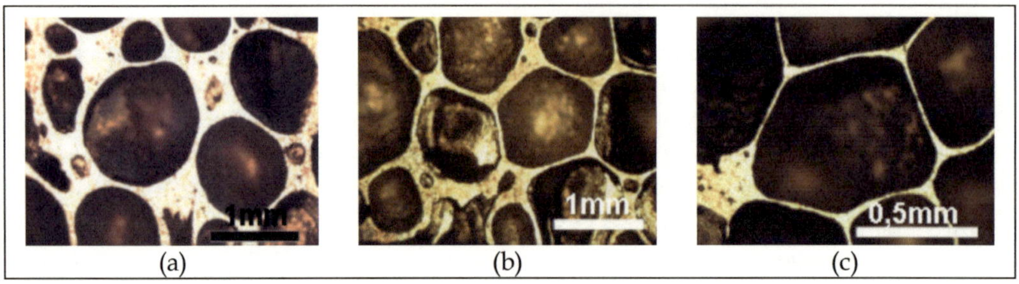

Fig. 21. Pore geometries: (a) spherical, (b) polyhedral and (c) other geometries.

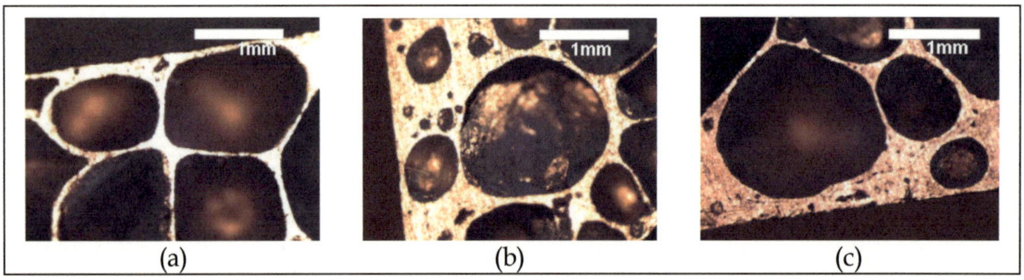

Fig. 22. Thicknesses of the dense surface skin in different sections a sample AlSi7 foam: (a) top, (b) lateral and (b) bottom sides.

4.2 Mechanical properties

Many literature studies have been undertaken on the mechanical properties of metal foams. A broad survey of the understanding of the mechanical behaviour of a wide range of cellular solids is provided by Gibson and Ashby (Gibson & Ashby, 2000). Others, have carried out experiments to investigate the behaviour of metallic foams under different loading conditions, particularly the properties of metal foams under impact loading. The possibility of controlling the load-displacement behaviour by an appropriate selection of matrix material, cellular geometry and relative density makes foams an ideal material for energy absorbing structures. Among the several mechanical testing methods available, uniaxial compressive mechanical tests are commonly used to evaluate the compressive behaviour and the energy absorbed of these foams. The elastic modulus, yield and plateau strengths are the most important mechanical properties parameters which are obtained from these curves. The stress-strain curves of closed-cell Al-alloy foams display either plastic or brittle fracture depending on foam fabrication and microstructure (Sugimura et al, 1997; Banhart & Baumeister, 1998).

The compression behaviour of these Al-alloy foams depends on several parameters such as: (i) the Al-alloy composition; (ii) the foam morphology (cell size range); (iii) the density gradient of samples; (iv) the defects of cellular structure (cell walls) and (v) the characteristics of the external surface skin. The influence of the density and the architecture of these foams on the mechanical properties are strong and complex. Depending on the material from which the foam is made, different mechanisms (brittle or ductile) can be observed. The compressive properties, such as, the average plateau stress, modulus, the elasticity and the energy absorption, depend, above all, on the foam density in which their values increase with the rise of the density (Fig. 23).

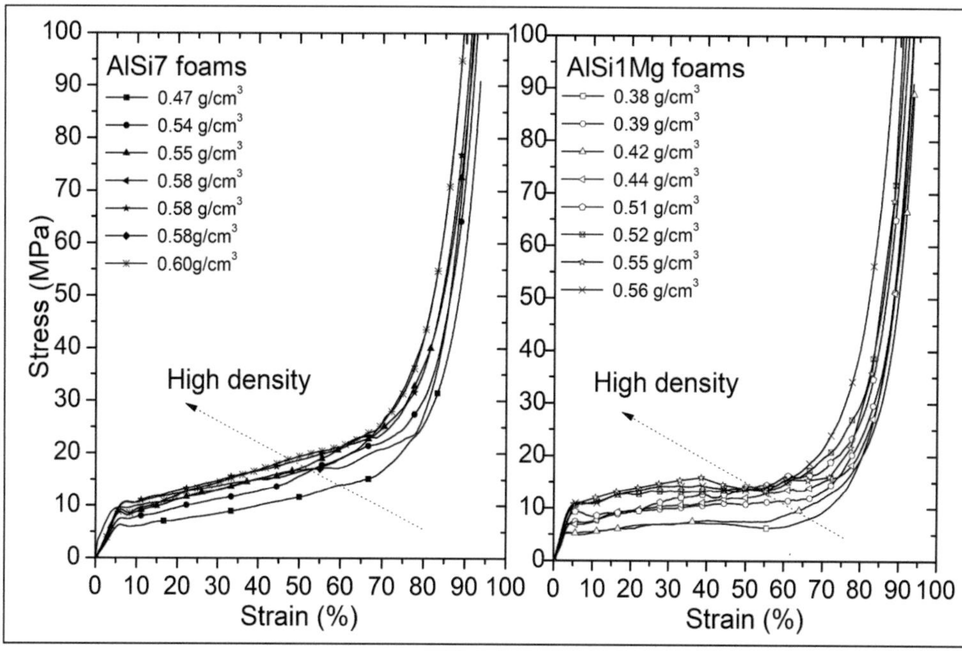

Fig. 23. Compressive stress-strain curves for the different specimens (cylindrical samples height/diameter ratio equal 1: h=ϕ= 30mm x 30mm).

As it can be depicted from Fig. 23, the stress-strain curves are divided into three characteristic regions. The first region (I) is linear-elastic where the load increases with increasing compression displacement almost linearly (elastic deflection of the pore walls), followed by a plastic collapse plateau (Region II) with a nearly steady compression load (pore walls yield or fracture, whereas increasing deformation does not require an increase of the load). The last region is the densification of the foam (Region III) where there is a rapid increase on the load after the cell walls crushed together.

These foams exhibit, after an elastic loading, a more or less clear plateau region. This plateau stress is important to characterise the energy absorbing behaviour and is a good material property for the compression performance of a foam. The measurements of the plateau stress depending on the different methods exist for measurement of the plateau stress depending on the course of the stress-strain curve.

The failure modes and mechanisms associated to these foams at different regions of the load-displacement have been identified (Fig. 24).

The elastic deformation occurs due to bending of the edges, elongation of cell walls and trapped gas pressure inside the cells. The deformation is not visible (point B, in Fig.24b). The latter is almost totally reversible, and occurs uniformly throughout the sample. After reaching the elastic limit, the collapse of the cells starts, mostly by distortion (stretching), rotation and/or sliding of the edges and cell walls, with permanent deformation (points C in Fig. 24b). A progressive collapse of the cells was observed in the "plateau" stage of the load-displacement curve (points C to F, in Fig. 24b). This deformation is not uniform due to the irregular structure of the foam (pore size distribution, thickness of cell walls, etc.) (Fig. 20a). The slope that characterises this region may be related to the compression of fluid trapped inside the cells, or due to tensile stress in the cell walls. The slope increases with increasing density of the foam. The shape of collapsed cells is very different from its original shape, as it contains bended and distorted cell walls that may even touch each other. However, in general, the fracture of cell walls does not occur. The initial collapse begins in a small group of cells in the region with the lowest local density of the sample. Collapse does not occur in all cells (points E, in Fig.24b), starting in the cells that are less resistant or with higher loads. The collapse of a cell induces the collapse of neighbouring cells. Moreover, the collapse of neighbouring cells evolves in successive layers and eventually leads to the formation of a single deformation band (points E in Fig.24b).

Al-alloy foams are often used as filler material in lightweight structures subject to crash and/or high velocity impact or as thermal/acoustic insulation devices. The energy

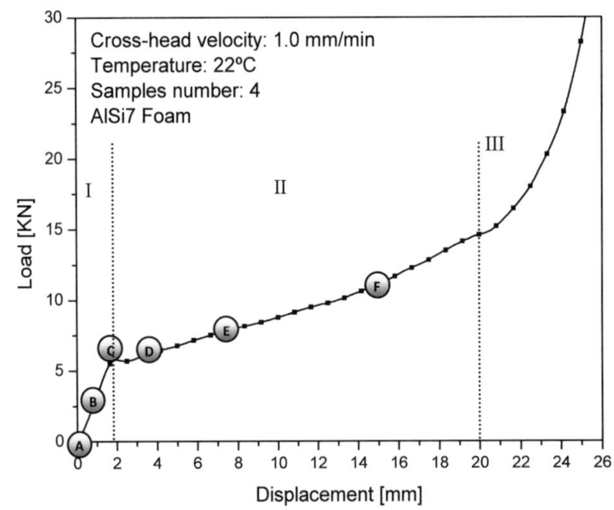

(a)

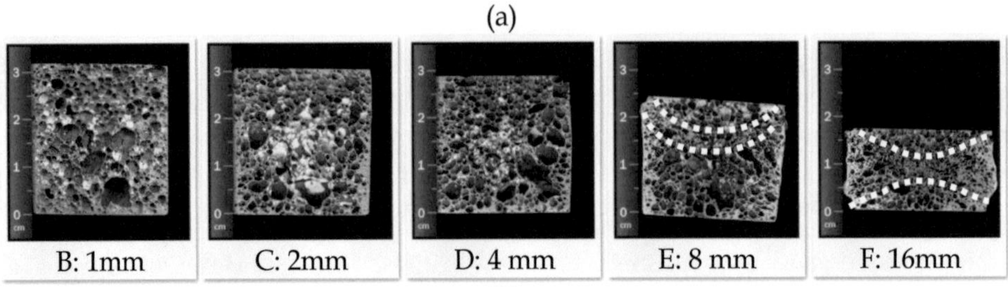

(b)

Fig. 24. Load-displacement curve of a AlSi7 foam under compressive loading. Cellular structures of deformed AlSi7 foam samples at different cross-head displacements.

absorption capability of these foams can be well estimated from the stress-strain compression behaviour of the material which is estimated from the area under the stress-strain curve (Fig. 25a). As foam materials exhibit a constant stress "plateau" they can absorb higher levels of energy than dense aluminium alloys. Most of the absorbed energy is irreversibly converted into a plastic deformation energy which is a further advantage of foamed Aluminium. For the same stress level, the dense material is deformed in the regime of reversible linear-elastic stresses, releasing most of the stored energy when the load is removed. Al-alloy foams exhibit higher energy absorption capabilities (Fig. 25). The increase of the energy absorption with increasing foam density is clearly obvious (Fig.25b).

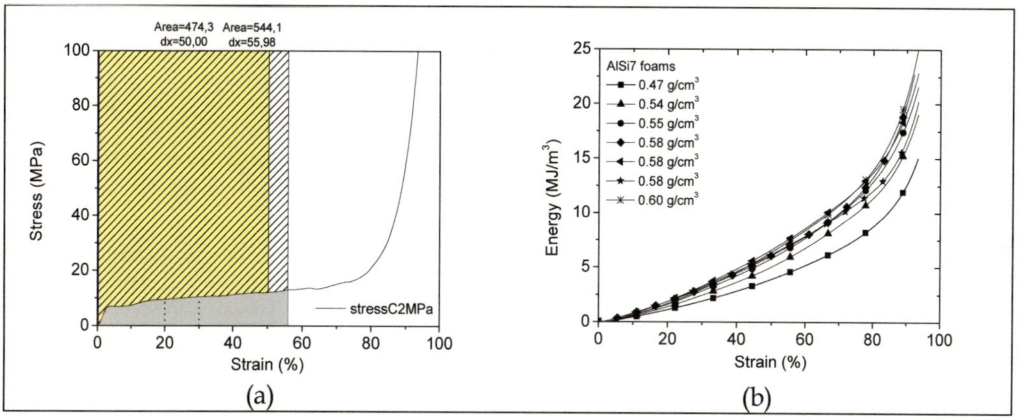

Fig. 25. (a) Absorbed energy per volume in a certain strain interval is the area under the stress-strain curve. (b) Absorbed energy curves of AlSi7 foam under compressive loading at different densities.

5. Challenges

Despite its technological advances, the metallic foam formation is not problem free and still poses challenges. Questions related to a very hot topic (i.e. the control of the pores size and shape of metal foams) that is seen with alacrity by the scientific community due to potential applications of these materials in the transport industry, are highlighted and discussed in detail. The key-question is how to produce metal foams, in series, achieving uniform cellular structure, in order to improve the manufacture reproducibility and to control foam architecture. A key goal of this group research work is to develop the missing knowledge to fill in the highlighted gap in the production of Al-alloy foams of uniform closed-cell structures and transfer it to industry.

6. References

Ashby, M.F., Evans, A., Fleck, N.A., Gibson, L.J., Hutchinson, J.W. & Wadley, H.N.G. (2000). *Metal foams – a design guide,* Butterworth-Heinemann, ISBN 0-7506-7219-6, London, England

Babcsán, N., García-Moreno, F. & Banhart, J. (2005). Metal foams—high temperature colloids: Part I. Ex situ analysis of metal foams. 261(1-3):123-130, 2005.

Babcsán, N., García-Moreno, F. & Banhart, J. (2007). Metal foams - High temperature colloids. Part II: In-situ analysis of metal foams. *Colloids and Surfaces A: Physicochemical and Engineering Aspects*, Vol. 309, No. (1-3), pp. 254–263, ISSN 0927-7757

Banhart, J. (2001). Manufacture, characterisation and application ofcellular metals and metal foams *Progress in Materials Science* Vol. 46, No. (6), pp. 559–632, ISSN 0079-6425

Banhart, J. (2006), Metal foams: production and stability, *Advanced Engineering Materials,* Vol. 8, No. (9), pp. 781–794, ISSN 1438-1656

Banhart, J. (2008). Gold and gold alloy foams. *Gold Bulletin,* Vol. 41, No. (3), pp. 251-256, ISSN 0017-1557

Banhart, J. & Baumeister, J. (1998). Deformation characteristics of metal foams. *Journal of Materials Science*, Vol. 33, pp. 1431–1440, ISSN: 0022-2461

Baumgärtner, F., Duarte, I. & Banhart, J. (2000). Industrialisation of P/M foaming process. *Advanced Engineering Materials*, Vol.2, No.4, pp. 168-174, ISSN 1438-1656

Belkessam, O. & Fritsching U. (2003). Modelling and simulation of continuous metal *foaming process. Modelling and Simulation in Materials Science and Engineering,* Vol. 11, No.6, pp. 823–837, ISSN 0965-03932003.

Brunke, O. & Odenbach, S. (2006). In situ observation and numerical calculations of the evolution of metallic foams. *Journal of Physics: Condensed Matter*, Vol. 18, pp. 6493–6506, ISSN 0953-8984

Campana, F. & Pilone, D. (2008). Effect of wall microstructure and morphometric parameters on the crush behaviour of Al-alloy foams. *Materials Science & Engineering A*, Vol. 479, No. (1-2), pp. 58-64, ISSN 0921-5093

Cambronero, L.E.G., Ruiz-Roman, J.M., Corpas, F.A. & Ruiz Prieto, J.M. (2009). Manufacturing of Al–Mg–Si alloy foam using calcium carbonate as foaming agent. *Journal of Materials Processing Technology*, Vol. 209, pp. 1803–1809, ISSN 0924-0136

Degischer, H.P. & Kriszt, B. (Ed(s).) (2002). *Handbook of Cellular Metals*, Wiley-VCH, ISBN 3-527-30339-1, Weinheim

Duarte, I. & Banhart, J. (2000). A study of aluminium foam formation - kinetics and microstructure. *Acta Materialia*, Vol.48, No.9, pp. 2349-2362, ISSN 1359-6454

Duarte, I., Banhart, J. Ferreira & M. Santos. (2006) Foaming around fastening elements. *Materials Science Forum*, Vol. 514-516, pp. 712-717, ISSN 0255-5476

Duarte, I., Santos, M. & Vide, M. (2008). Processo contínuo de produção de peças e protótipos em espumas metálicas. Ingenium, II série, Março/Abril 2008, Vol. 104, pp. 78-80, ISSN 0870-5968I.

Duarte, I. (2005), Espumas Metálicas: Processos de fabrico, Caracterização e Simulação numérica. PhD thesis. FEUP- Faculty of Engineering of University of Porto, Porto.

Duarte, I., Teixeira-Dias, F., Graça, A. & Ferreira, AJM. (2010). Failure Modes and Influence of the Quasi-static Deformation Rate on the Mechanical Behavior of Sandwich Panels with Aluminum Foam Cores. *Mechanics of Advanced Materials and Structures*, Vol.17, No. (5), pp. 35-342, ISSN 1537-6494

Fang, J., Yang., Z., Zhang, H. &Ding, B. (2005). The coating process of silica film on TiH2 particles and gas release characteristic. *Chemical Engineering Science*, Vol. 60, pp. 845-850, ISSN 0009-2509

Garcia-Moreno, F. , Babcsan, N. & Banhart, J. (2005). X-ray radioscopy of liquid metal foams: influence of heating profile, atmosphere and pressure. *Colloids and Surfaces A: Physicochemical and Engineering Aspects*, Vol. 263, pp. 290–294, ISSN 0927-7757

Gibson, L.J. & Ashby, M.F. (1997). *Cellular solids – Structure and properties*, Second Edition, Cambridge University Press, ISBN 0-521-49911-9, Cambridge, United Kindgom

Gokhale, A., Sahu, S.N., Kulkarni, V.K.W.R., Sudhakar, B., Rao & N. Ramachandra (2007) Effect of Titanium Hydride Powder Characteristics and Aluminium Alloy Composition on Foaming. *High Temperature Materials and Processes*. Vol. 26, No.4, pp. 247–256, ISSN 0334-6455

Haesche, M., Lehmhus,D., Weise, J., Wichmann, M. & Mocellin, I. C.M. (2010). Carbonates as Foaming Agent in Chip-based Aluminium Foam Precursor. *Journal of Materials Science & Technology*, Vol. 26, No. (9), pp. 845-850, ISSN 1005-0302

Helwig, H.-M., Garcia-Moreno, F. & Banhart, J. (2011). A study of Mg and Cu additions on the foaming behaviour of Al–Si alloys. *Journal of Materials Science*, Vol. 46, No. (15), pp. 5227-5236, ISSN 0022-2461.

Ibrahim, A., Körner, C. & Singer, R.F. (2008). The effect of TiH2 particle size on the morphology of Al-foam produced by PM process. *Advanced Engineering Materials*, Vol. 10, pp. 845–848, ISSN 1438-1656

Lehmhus, D. & Busse, M. (2004). Potential New Matrix Alloys for Production of PM Aluminium Foams. *Advanced Engineering Materials*, Vol. 6, pp. 391-396, ISSN 1438-1656.

Jiménez, C., García-Moreno, F., Mukherjee, M., Görke, O., Banhart, J. (2009). Improvement of aluminium foaming by powder consolidation under vacuum. *Scripta Materialia*, Vol.61, No.5, pp., 552–555, ISSN 1359-6462

Markaki, A.E. & Clyne, T.W. (2001). The effect of cell wall microstructure on the deformation and fracture of aluminum-based foams, *Acta Materialia*, Vol. 49, N.o 9, pp. 1677-1686, ISSN 1359-6454

Matijasevic, B. & Banhart, J. (2006). Improvement of aluminium foam technology by tailoring of blowing agent. *Scripta Materialia*, Vol. 54, No. (4), pp. 503-508, ISSN 1359-6462

Matijašević, B., Banhart, J., Fiechter, S., Görke, O., Wanderka, N. (2006), Modification of titanium hydride for improved aluminium foam manufacture, *Acta Materialia*, Vol. 54, N.o 7,pp. 1887–1900, ISSN 1359-6454

Miyoshi, T., Itoh, M., A. kiyama, S. & Kitahara, A. (2000). Alporas aluminum foam: production process, properties, and applications. *Advanced Engineering Materials*, Vol. 2, No. 4, pp. 179-183, ISSN 1438-1656

Mukherjee, M (2009). *Evolution of metal foams during solidification, Technischen Universität Berlin*

Proa-Flores, P.M. & Drew, RAL, (2008).Production of Aluminum Foams with Ni-coated TiH2 Powder. *Advanced Engineering Materials*, Vol. 10, No. (9), pp. 830-834, ISSN 1438-1656

Kennedy, A. R. (2002). Effect of compaction density on foamability of Al-TiH2 powder compacts. *Powder Metallurgy*, Vol. 45, No. (1), pp. 75-79, ISSN 0032-5899

Kennedy, A. R. (2004). Effect of foaming configuration on expansion. *Journal of Materials Science*. Vol. 39, pp. 1143-1145, ISSN 0022-2461

Kitazono, K., Sato, E. & Kuribayashi, K. (2004), Novel manufacturing process of closed-cell aluminum foam by accumulative roll-bonding. Scripta Materialia, Vol. 50, No. (4), pp. 495–498, ISSN 1359-6462

Kolluri, M., Mukherjee, M., Garcia-Moreno, F., Banhart, J. & Ramamurty, U. (2008). Fatigue of a laterally constrained closed cell aluminum foam. Acta Materialia. Vol. 56, pp. 1114, ISSN 1359-6454

Körner, C., Thies, M. & Singer, R. F. (2002). Modeling of Metal Foaming with Lattice Boltzmann Automata. *Advanced Engineering Materials,* Vol. 4, N.o 10, pp. 765 - 769, ISSN 1438-1656

Körner, C., Arnold, & Singer, R. F. (2005). Metal foam stabilization by oxide network particles. *Materials Science and Engineering A,* Vol. 396, N.o (1-2), pp. 28–40, ISSN 0921-5093.

Körner, C. (2008). Foam formation mechanisms in particle suspensions applied to metal foams. *Materials Science and Engineering A,* Vol. 495, N.o (1-2), pp. 227-235, ISSN 0921-5093.

Rack, A., Helwig, H.-M., Bütow, A., Rueda, A., Matijasevic´-Lux, B., Helfen, L., Goebbels, J. & Banhart, J. (2009). Early pore formation in aluminium foams studied by synchrotron-based microtomography and 3-D image analysis. *Acta Materialia* 57(16): 4809–4821, 2009, ISSN 1359-6454

Ramamurty, U. & Paul, A. (2004). Variability in mechanical properties of a metal foam, *Acta Materialia,* Vol. 52, No. (4), pp. 869-876, ISSN 1359-6454

Sugimura, Y., Meyer, J., He, M.Y., Bart-Smith, H., Grenstedt, J. & Evans, A.G. (1997). On the mechanical performance of closed cell Al alloy foams. *Acta Materialia,* Vol. 45, No. (12), pp.5245-5259, ISSN 1359-6454

Stanzick, H., Helfen, L., Danikin, S. & Banhart, J. (2002). Material flow in metal foams studied by neutron radioscopy. *Applied Physics A: Materials Science & Processing,* Vol. 74, N.o (1), pp. 1118-1120, ISSN 0947-8396

Stavans, J. (1993). The evolution of cellular structures. *Reports on Progress in Physics,*Vol. 56, pp. 733-89, ISSN 0034-4885

Zeppelin, F., Hirscher, M., Stanzick, H. & Banhart, J (2003). Desorption of hydrogen from blowing agents used for foaming metals. *Composite Science and Technology,* Vol. 63, pp. 2293–2300, ISSN: 0266-3538

4

The Fabrication of Porous Barium Titanate Ceramics via Pore-Forming Agents (PFAs) for Thermistor and Sensor Applications

Burcu Ertuğ
Gedik University,
Department Of Metallurgical & Materials Engineering, Yakacık/Kartal, Istanbul
Turkey

1. Introduction

The perovskite family includes many titanates used in various electroceramic applications, for example, electronic, electro-optical, and electromechanical applications of ceramics. Barium titanate, perovskite structure, is a common ferroelectric material with a high dielectric constant, widely utilized to manufacture electronic components such as mutilayer capacitors (MLCs), PTC thermistors, piezoelectric transducers, and a variety of electro-optic devices (Wang, 2002).

For positive temperature coefficient of resistance (PTCR) and humidity/gas sensing applications of $BaTiO_3$, a porous microstructure is required. When oxygen is adsorbed on the grain boundaries, PTCR effect is enhanced. Thus a porous structure is needed for PTCR properties. Besides, porous ceramics can also be used for humidity/gas sensing. The water vapour is adsorbed on pores improving the electrical conductivity of the surface.

A number of routes are employed in order to fabricate porous ceramics for PTCR and/or humidity/gas sensors. Low pressure forming can be used for the production of $BaTiO_3$ ceramics. Forming can be done by unaxial pressing since hot pressing or hot isostatic pressing can eliminate pores. Also several other pressureless forming techniques can be used to make porous bodies. Another method for fabricating porous ceramics is using pore forming agents (PFAs) prior to sintering. PFAs form porosity through the ceramic body by different mechanisms. Different porosifiers that can be used for the production of barium titanate ceramics are starch based, carbon based, metallic or polymeric porosifiers. In the following paragraphs, crystal structure of, donor doping of and the electrical properties of barium titanate is described. Also the concept of porosity is briefly mentioned. Porous ceramics production, pore formers in general and pore formers in $BaTiO_3$ is also explained. Finally PTCR and humidity/gas sensing properties of $BaTiO_3$ is described in detail.

2. Perovskites and barium titanate ($BaTiO_3$)

ABO_3 oxides which adopt the perovskite structure form an important group of compounds possessing many useful and interesting physical properties. The idealized structure is cubic, point group m3m. The structure can be thought of in two ways. First, it can be viewed as being constructed from corner-sharing octahedral of oxygen ions, with the small B cations occupying the octahedral site and the much larger A cations occupying the interstices between the stacked octahedra. Alternatively, the structure can be thought of as close-packed ordered layers of oxygen and the larger A cations (AO_3), with the B cations occupying the octahedral interstices between the layers (Brook, 1991). The perovskite lattice of barium titanate is shown in Fig.2.1.

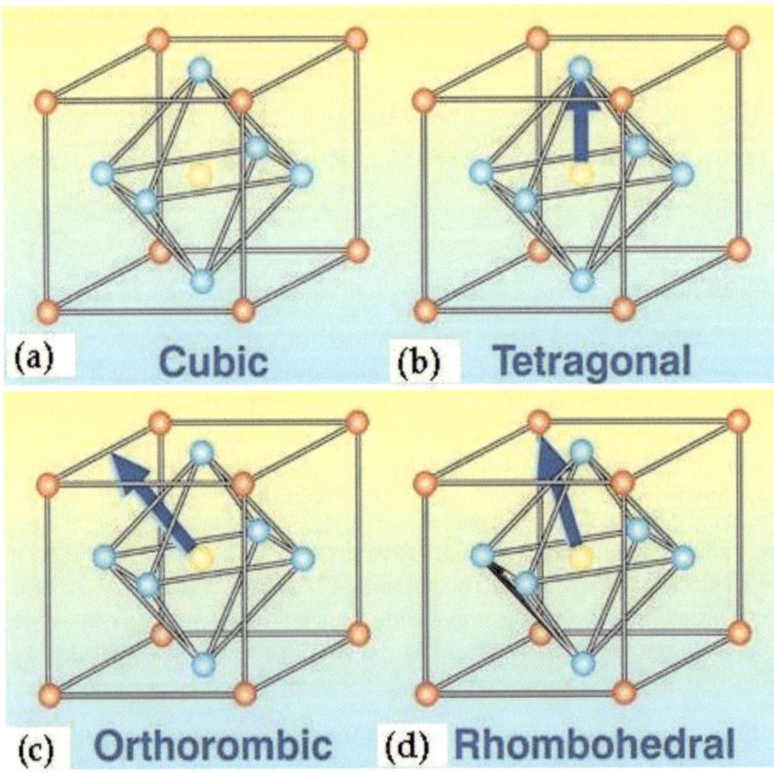

Fig. 2.1 The crystal structures of barium titanate ($BaTiO_3$). (b) [001], (c) [110], and (d) [111] directions (Hutagalung, 2011).

$BaTiO_3$ exhibits two basic structures, one being a ferroelectric, tetragonal, perovskite structure within temperature range 5-125°C. Above 125°C, the $BaTiO_3$ has the cubic perovskite, paraelectric phase (a=4.009A°, c/a=1). BaTiO3 exhibits a first-order displacive phase transition at 125°C from cubic (m3m) to tetragonal (4mm) on cooling. This change is accompanied by an elongation in the [001] direction and a contraction along the a-axes. Below 5°C, $BaTiO_3$ has another displacive phase transition resulting in an orthorhombic (mm2) structure (a=5.667A°, b=5.681A°, c=3.989A°). Associated with the phase change, there is a continuous change in the polarization direction, from parallel to the <100> direction of the unit cell, towards the <110> direction. At -90°C, $BaTiO_3$ is transformed into the rhombohedral crystal structure and the polarization direction becomes parallel to the <111> direction (Zhou, 2000). The change of the crystal lattice dimensions with temperature in $BaTiO_3$ is shown in Fig.2.2.

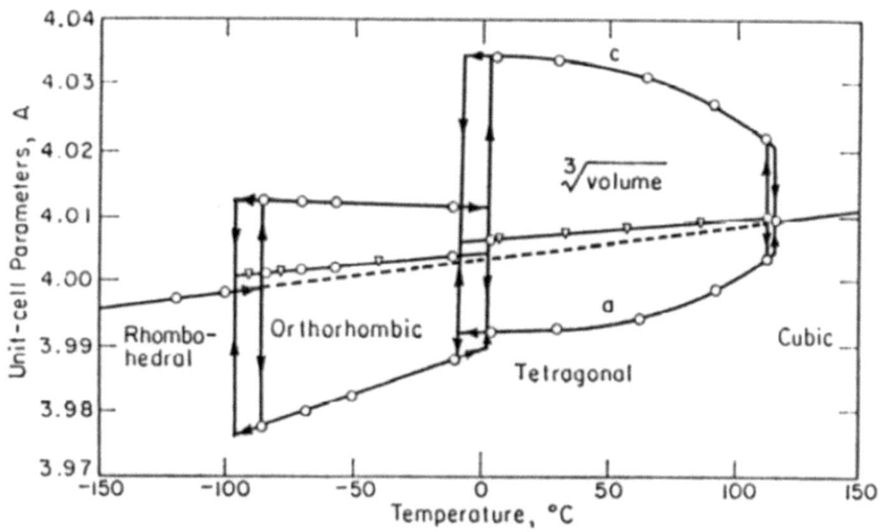

Fig. 2.2 Lattice constants of $BaTiO_3$ as function of temperature (Wang, 2002).

In an ideal cubic perovskite, the ionic radii, r_i (i=A, B, O), satisfy the relation $r_A + r_O = \sqrt{2}\,(r_B + r_O)$. The Goldschmidt tolerance factor for perovskites is, hence defined by

$$t = \{r_A + r_O\} \,/\, \{\sqrt{2}\,(r_B + r_O)\} \quad (2\text{-}1)$$

For pure barium titanate t~1.06. t is greater than 1 due to the fact that the Ti^{4+} ion is smaller than its cavity ($BaZrO_3$ has t~1) and/or that the Ba^{2+} is larger than its cavity ($SrTiO_3$ has t~1) (Tsur et al., 2001).

2.1 Donor doping of $BaTiO_3$

In a stoichiometric solid solution the extra positive charge of a donor center can be compensated by a cation vacancy or an anion interstitial

$$D_2O_3(2MO) = 2D^{\cdot}_M + O_O + O''_I \quad (2\text{-}2)$$

$$D_2O_3(3MO) = 2D^{\cdot}_M + V''_M + 3O_O \quad (2\text{-}3)$$

Eq. (2-2) maintains a perfect cation sublattice, while the anion sublattice is perfect in Eq.(2-3). A shift from the stoichiometric solid solution with its compensation by a lattice defect, to a non-stoichiometric solid solution with compensation by an electronic defect, requires an interaction with the ambient atmosphere. The stoichiometric solution may gain or lose oxygen at oxygen activities greater than or less than, in equilibrium with the stoichiometric composition. In this way, The compensating lattice defect is eliminated and is replaced by electrons in the case of donor dopants (Smyth, 2000).

When La^{3+} replaces Ba^{2+} on the A-site (La is too large to replace Ti on the B-site), charge imbalance is created which must be compensated by either cation vacancies on the A- or B-site (ionic compensation), or by electrons (electronic compensation) as follows

$$2La_2O_3 + 3TiO_2 = 4La^{\cdot}_{Ba} + 3Ti_{Ti}^{x} + 12O_O^{x} + V_{Ti}'''' \quad (2\text{-}4)$$

$$2La_2O_3 + 3TiO_2 = 2La^{\cdot}_{Ba} + 3Ti_{Ti}^{x} + 9O_O^{x} + V_{Ba}'' \quad (2\text{-}5)$$

$$2La_2O_3 + 2TiO_2 = 2La^{\cdot}_{Ba} + 2Ti_{Ti}^{x} + 6O_O^{x} + 1/2O_2 + 2e' \quad (2\text{-}6)$$

Ionic compensation ((2-4) and (2-5)) should have negligible effect on the room temperature conductivity due to the immobility of cation vacancies; La doped $BaTiO_3$ compensated in this way should, therefore, remain insulating. In contrast, electronic compensation (Eq. (2-6)) should cause a substantial increase in conductivity, in which the number of carriers equals the La concentration (Morrison et al., 2001).

Fig.2.3 indicates the effect of donor concentration (La^{3+}) on the electrical conductivity and grain size of $BaTiO_3$. As the donor concentration increases initially the conductivity increases up to 0.15% of La^{3+} and then decreases up to 0.3% of La^{3+}. The high conductivity region is where the electronic compensation dominates, after a critical donor concentration, cation vacancy compensation dominates and electrical conductivity decreases. Up to 0.3% of La^{3+}, the grain size of $BaTiO_3$ is not affected by donor concentration being 25 μm, however, above 0.3% of La^{3+}, grain size decreases to 5μm.

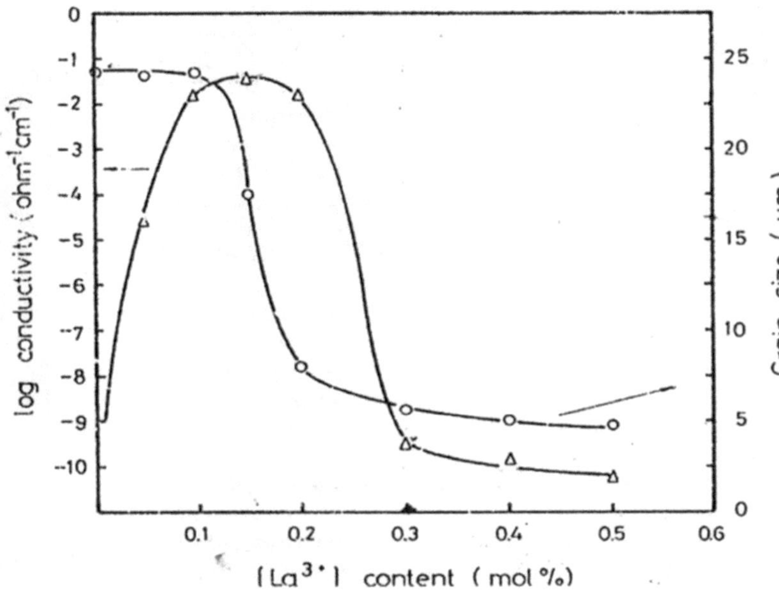

Fig. 2.3 Schematic of donor concentration influence on room temperature electrical conductivity and grain size.

Donor dopant incorporation is achieved by either electronic compensation at low concentrations or vacancy compensation at high concentrations. High concentrations of segregating donors at grain boundaries inhibites grain growth. At small concentrations, donor incorporation by electronic compensation explains the high conductivity. As the average dopant concentration increases, the local donor concentration at the grain boundary increases rapidly due to segregation. The donor incorporation at the grain boundary shifts from electronic to vacancy compensation, resulting in the formation of highly resistive layers and also, grain size decreases due to significant dopant drag on the boundary mobility (Desu, 1990).

3. Positive temperature coefficient of resistance (PTCR) property

Barium titanate is a very attractive material in the field of electroceramics and microelectronics due to its good characteristics. Its high dielectric constant and low loss characteristics make barium titanate an excellent choice for many applications, such as capacitors, multilayer capacitors (MLCs) and energy storage devices. Doped barium titanate has found wide application in semiconductors, positive temperature coefficient resistors, ultrasonic transducers, piezoelectric devices, and has become one of the most important ferroelectric ceramics (Vijatovic et al., 2008).

Barium titanate ($BaTiO_3$) is an insulator with a large energy gap of 3.05 eV at room temperature. It can be made n-type semiconducting by partially substituting Ba^{2+} by trivalent cations such as La^{3+} and Y^{3+} or Ti^{4+} by pentavalent cations such as Nb^{5+} and Sb^{5+}. The extra positive charge is compensated by free electrons. Doped $BaTiO_3$ which is sintered in air, exhibits a very high relative permittivity and a sharp increase in resistivity above Curie temperature, i.e 120°C (called "positive temperature coefficient of resistance", or PTCR) when the structure changes from ferroelectric to paraelectric. Pure barium titanate is an insulator with no PTCR effect. A high PTCR effect can be achieved by the incorporation of transition metals such as Mn, V, Fe, Cu and Cr. This is because the transition metals substitute for Ti $^{4+}$ as a lower valency state and act as acceptors (Kim, 2002; Kim et al., 2000; Park, 2004).

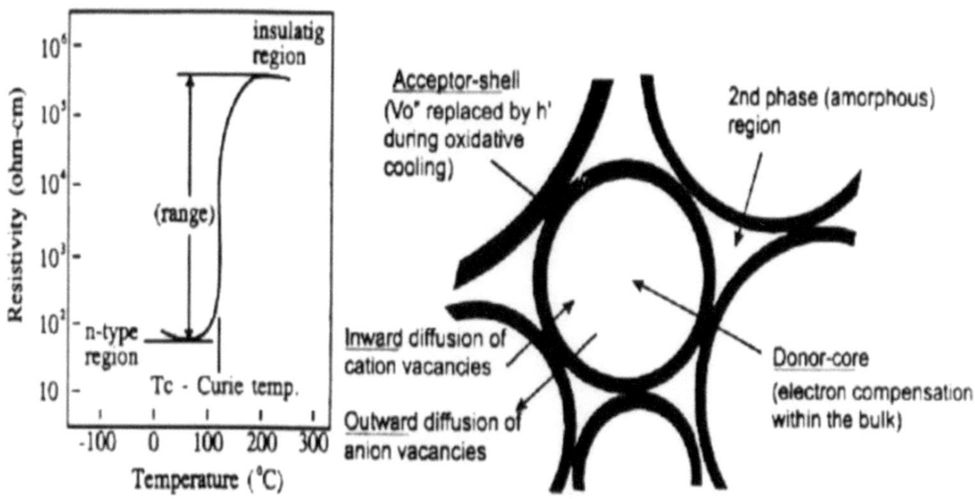

Fig. 3.1 Electrical resistivity for typical PTCR device and schematic presentation of defect chemistry responsible for PTCR effect (Vijatovic et al., 2008).

The grain-boundary model given by Heywang treats the grain boundary as a n-type Shottky barrier with deep acceptor states at grain boundaries (Vijatovic et al., 2008). The resistance anomaly behavior of doped BaTiO3 is shown in Fig.3.1.

The double-Schottky energy barrier model proposed by Heywang is one of the most effective model which explains the mechanism of PTCR effect. According to this model, the double-Schottky type energy barrier is formed along the grain boundary and its barrier height determines the total conductivity. The electrical potential barrier results from segregated acceptor ions or adsorbed oxygen at the grain boundaries. Sintering under vacuum or reducing gas atmosphere strongly reduces PTCR effect (Zhang, 2004).

Heywang explained PTCR effect in terms of potential barriers. The double Schottky barriers form at the grain boundaries. These barriers result from electron trapping by acceptor states at the interfaces. The acceptor states arise from segregated impurities at the grain boundaries or from adsorbed oxygen or from barium vacancies according to Heywang, Jonker and Daniels, respectively. Below Curie temperature in tetragonal $BaTiO_3$, barium vacancies are neutral but above Curie temperature vacancies trap electrons and become activated to singly ionised barium vacancies. Thus a negatively charged region forms at the grain boundary and a positively charged space region forms adjacent to the boundaries. A potential barrier is created between these two regions. Since trapped electrons are not available for conduction, the electrical resistivity increases around Curie temperature (Steele, 1991).

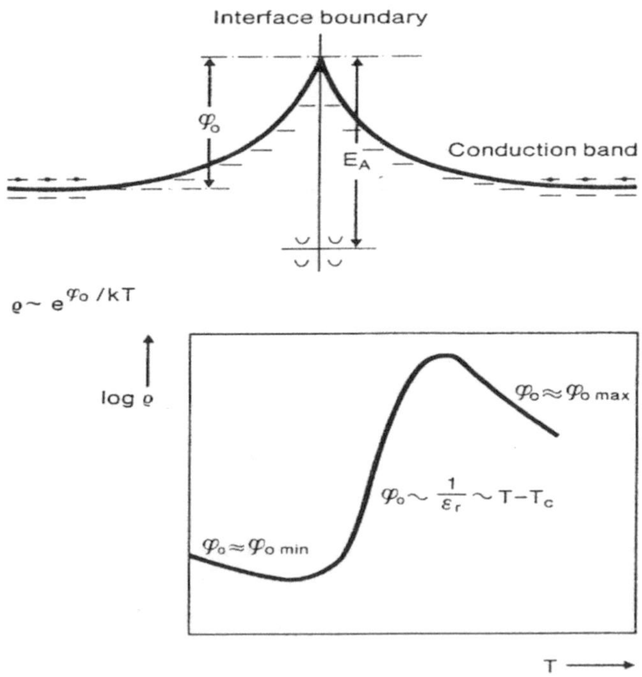

Fig. 3.2 The height of the potential barrier and the electrical resistivity-temperature curve (Steele, 1991).

The height of the potential barrier, φ_0 (shown in Fig.3.2) is given by

$$\varphi_0 = A / \varepsilon = A / C . (T\text{-}Tc) \quad (3\text{-}1)$$

where A is a constant, ε is permittivity and Tc is the Curie temperature. Above Tc, the height of the potential barrier, φ_0 increases with decreasing ε (Steele, 1991). The electrical resistivity, ρ is directly proportional to the height of the potential barrier according to

$$\rho \sim \exp (\varphi_0 / k.T) \quad (3\text{-}2)$$

Thus the exponential increase of the electrical resistivity, ρ is

$$\rho \sim \exp A / C.k\,(1 - Tc / T) \qquad (3\text{-}3)$$

The barrier height of $BaTiO_3$ ceramics increases with increasing porosity. There is an exponential relationship between the resistance of PTCR ceramics and the barrier height. If the ferroelectric compensation of the ceramics at the room temperature is the same, the resistance of PTCR ceramics increases with increasing porosity. With the increase of porosity, the contact between individual grains looks more dispersed, which is helpful to diffusion of oxygen into the ceramic bulk and to oxidation of grain boundaries. A large PTCR jump is only possible for the samples sintered in an oxidizing atmosphere. The oxygen concentration at grain boundary is in direct proportion to PTCR effect of ceramics (Zhang, 2004).

Sintering under O_2 atmosphere affects not only barrier height but the resistance and capacitance of grain boundaries. The effect is due to variation of the adsorbed gases at the grain boundaries. The diffusion coefficient of oxygen at grain boundary is much higher than that in bulk, which implies that there is a gradient of adsorbed oxygen density in the surface of the grain. Sintering under O_2 atmosphere results in increasing number of oxygen acceptors at the grain boundary and increase in resistance. The barrier height also increases with sintering (Jiang, 2003).

3.1 Barium titanate as PTCR thermistor

Barium titanate, which has a perovskite structure, is a common ferroelectric material with a high dielectric constant, widely utilized to manufacture electronic components such as mutilayer capacitors (MLCs), PTC thermistors, piezoelectric transducers, and a variety of electro-optic devices. Dielectrics and insulators can be defined as materials with high electrical resistivities. A good dielectric is, of course, necessarily a good insulator, but the converse is by no means true (Wang 2002). Barium titanate ($BaTiO_3$) is a ceramic material with a band gap of 3.05 eV at room temperature. Since the energy gap of barium titanate is large, it is an insulator when it is pure. It can be made n-type semiconductor when doped with donor dopants such as La^{3+} and Nb^{5+}. Poly-crystalline *n*-type semiconducting barium titanate ($BaTiO_3$) exhibits a behaviour known as the positive temperature coefficient of resistivity (PTCR) effect. The electrical resistivity of *n*-type semiconducting barium titanate increases by several orders of magnitude near the ferroelectric Curie temperature (120 °C) (Kim 2000). At the Curie temperature, barium titanate undergoes ferroelectric to paraelectric transition (Kim 2004). This behaviour is indicated in Fig. 3.3 (Wang 2002). It has also been reported that single crystals of barium titanate exhibit negative temperature co-efficient of resistivity (NTCR) properties.

In fact, PTCR materials can be divided into four groups: polymer composites, ceramic composites, V_2O_3 compounds and BaTiO3-based compounds ($BaSrTiO_3$, $BaPbTiO_3$...) (Wang 2002). Theoretical models of the PTCR effect have been proposed by both Heywang and Jonker. Both theories and experiments report that the effect is a grain-boundary related phenomenon (Park 2002). The grain boundaries of barium titanate show a large electrical resistance and PTCR results from the adsorbed oxygen at the grain boundaries. The porous ceramics exhibit large PTCR effect. According to Heywang, PTCR effect can be explained in

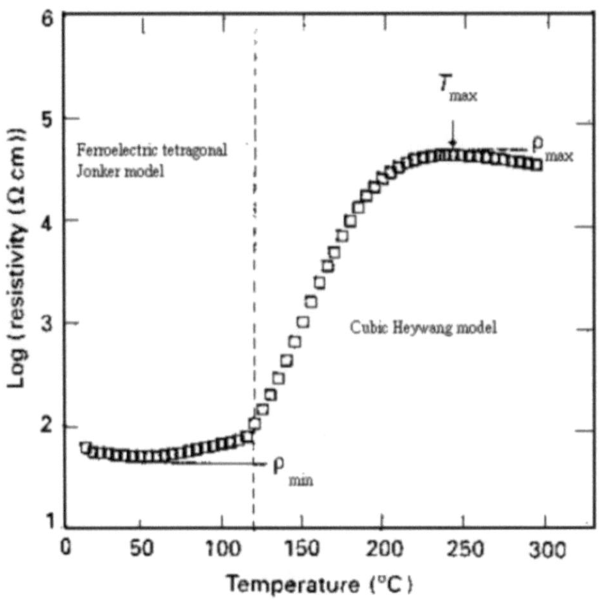

Fig. 3.3 Typical resistivity behavior of a BaTiO3 –type PTCR material (Wang 2002).

terms of temperature-dependent Schottky-type grain boundary potential barriers. These potential barriers result from charge accumulation due to grain boundary acceptor states such as segregated acceptor ions or adsorbed oxygen at the grain boundaries.

When the pure $BaTiO_3$ is sintered in air, the defect reaction is as follows:

$$1/2O_2(g) = O_O^x + V_{Ba}^x \qquad (3\text{-}4)$$

Because of the high oxygen pressure at grain boundary when samples are sintered in air, this reaction prefers to happen at the grain boundary rather than in the grain lattice.The neutral barium vacancy may be ionized by the electron that was introduced by the donor dopant.

$$V_{Ba}^x + 2e- = V_{Ba}'' \qquad (3\text{-}5)$$

or

$$D_2O_3 + V_{Ba}^x = 2D^{\cdot}_{Ba} + V_{Ba}'' + 2O_O^x + 1/2O_2(g) \qquad (3\text{-}6)$$

In pure donor doped $BaTiO_3$ ceramics (the microstructure is given in Fig. 3.4), the capturing of a free electron by neutral barium vacancy at the grain boundary during ferroelectric phase transition would respond for the PTCR effect (Qi, 2002).

PTCR effect can only be observed when barium titanate is sintered in air or oxidizing atmospheres. However, when it is sintered under vacuum or in reducing gas atmospheres, magnitude of the resistivity jump is reduced critically(Kim 2004). While the samples were annealed in reduced air (98% N_2 + 2% H_2), anion vacancies (oxygen vacancies) were generated and electron compensation would occur to maintain the electrical neutrality, increasing the conductivity of the ceramics. Heywang considered that no oxygen was absorbed in the grain boundaries to form a depletion layer, and therefore that vacancy compensation (Ba compensation) was suppressed and the potential barrier was not built, the ceramic exhibited a poor PTCR effect. As the oxygen pressure increases, foreign ions are increasingly compensated by the oxidation process (formation of barium vacancies), and this Schottky barrier was formed between grains and grain boundaries to benefit the PTCR effect (Chen 2007). At low temperatures resistance is low since resistance is dominated by the grains. Above Curie temperature the resistivity of grain boundary increases PTC resistance. The disappearance of the domain structure above the tetragonal-cubic/ferroelectric-paraelectric transition temperature (Tc) results in the energy barrier being raised at all the grain boundaries (Wegmann 2007). Positive temperature coefficient (PTC) materials prepared from doped semiconducting barium titanate ceramics can be used in various kind of electronic circuitry as a switching device or as a constant temperature heater (Wang 2002).

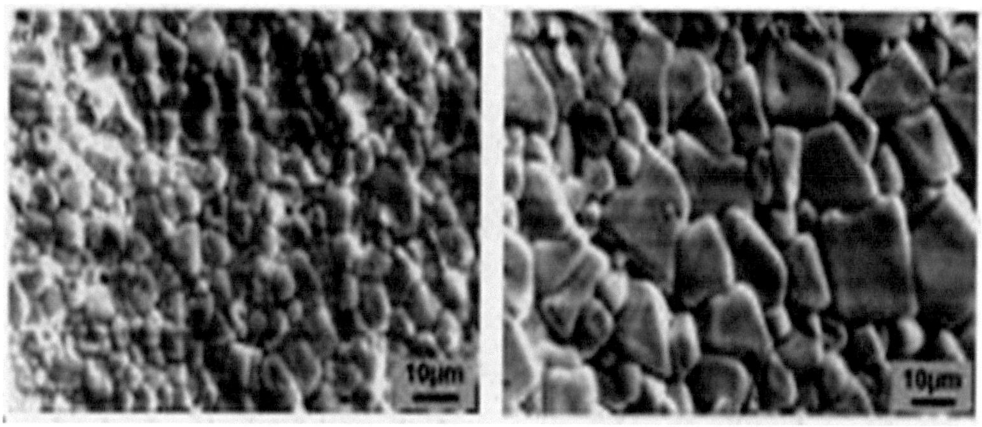

Fig. 3.4 The microstructure of BaTiO3 based PTCR materials sintered at (a) 800C and (b) 1150C.

These PTC materials prepared from doped semiconducting $BaTiO_3$ ceramics can be used in various kinds of electronic circuitry as a switching device or as a constant temperature heater. Other important application of a PTC thermistor is the measurement/detection/ control of temperature or parameters related to temperature. These PTC materials are known to have the highest temperature coefficient of resistance among all sensor materials available (Vijatovic et al., 2008).

4. Humidity/gas sensing

Even though the use of sensors is well established in process industries, agriculture, medicine, and many other areas, the development of new sensing materials with high sensing capabilities is proceeding at an unprecedented rate. Numerous materials have been utilized for humidity sensing, of which the metal oxides that are physically and chemically stable have been extensively investigated at both room and elevated temperatures. Sensors based on changes in resistance and capacitance, as shown in Fig.4.1, are preferred owing to their compact size, which could facilitate miniaturization required for electronic circuitry (Pokhrel et al., 2003).

These PTC materials are known to have a high temperature coefficient of resistance around Curie point and the ability of self-limiting, so they are as well come out to be a very useful device for sensor applications. The first semiconductor oxide gas sensors were reported by Seiyama et al. in 1962. Since then, there have been numerous studies concerning such oxide semiconductors as SnO_2, ZnO_2, In_2O_3, TiO_2, Fe_2O_3, HfO_2, and $BaSnO_3$. They are nowadays widely used in the detection of gases. Semiconductor oxide gas sensors are extensively studied in order to improve their sensing characteristics, i.e., sensitivity, selectivity, stability, and response rate, to various kinds of gases and to meet the increasing needs of sensors in complicated systems and under strict conditions. A trial-and-error method is still mainly used in the development of new sensor materials to replace existing sensor materials (Park 2004). Metal-oxide-semiconductor (MOS) sensors operate on the basis of the modification of electrical conductivity of metal oxide layers, resulting from the interactions between O^{2-}, O^- and O^{2-} species and gas molecules to be detected (Kim 2005).

It has been known that absorption or desorption of a gas on the surface of a metal oxide changes the conductivity of the material. The sensitivity of a surface to a gas can be as low as parts per billion (ppb). It is highly desirable that metal oxide semiconductor sensors have a large surface area, so as to adsorb as much of the target analyte as possible on the surface, giving a stronger and more measurable response (especially at low concentrations).

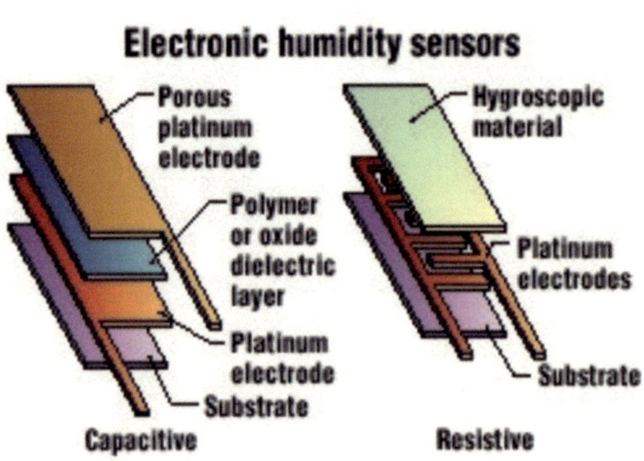

Fig. 4.1 Capacitive and resistive humidity sensors (Electronic Circuits and Projects Forum).

For n-type semiconductors, the majority charge carriers are electrons, and upon interaction with a reducing gas an increase in conductivity occurs. Conversely, an oxidising gas serves to deplete the sensing layer of charge carrying electrons, resulting in a decrease in conductivity (Fine 2010). Barium titanate is an n-type semiconductor which has humidity and gas (such as reducing gases) sensing properties (Park 2004, Wagiran 2005).

In recent years, solid state humidity sensors have been widely used for the measurement and control of humidity in an industrial or household environment. The relative humidity (RH), which is the ratio of actual vapour pressure of water to the saturated vapour pressure at a particular temperature, is commonly used to measure humidity. The sensor materials used are in general polymeric or ceramic. Ceramic humidity sensors are superior in performance to the polymeric type because of ist stability towards a variety of chemical species, wide range of operating temperatures and fast response to the changes of humidity. The principle of humidity sensing by ceramic sensors is changes of electrical conduction or capacitance due to water chemisorption and/or capillary conduction in the pores. The important ceramic materials used for humidity sensing are TiO_2, $Mg_2Al_2O_4$, hematite, ZnC_2O_4, perovskites. Perovskites such as (Ba, Sr) TiO_3 are protonic ionic conductors which have also been used for humidity sensing. In perovskite type oxides (ABO_3), the atoms on site A are susceptible to humidity (Agarwal and Sharma, 2002).

Perovskite oxides form a group of ceramics demonstrating a variety of interesting properties and promising applications. Some perovskite oxides, which are n-type semiconductors, such as doped $SrTiO_3$ ($Sr_xLa_{1-x}TiO_3$), exhibit dependence of conductivity on humidity in the atmosphere they are exposed to. It has been proposed that H_2O molecules adsorb on the surface, and the adsorbed H_2O molecules can release electrons to the conduction band and consequently increase the electronic conductivity.

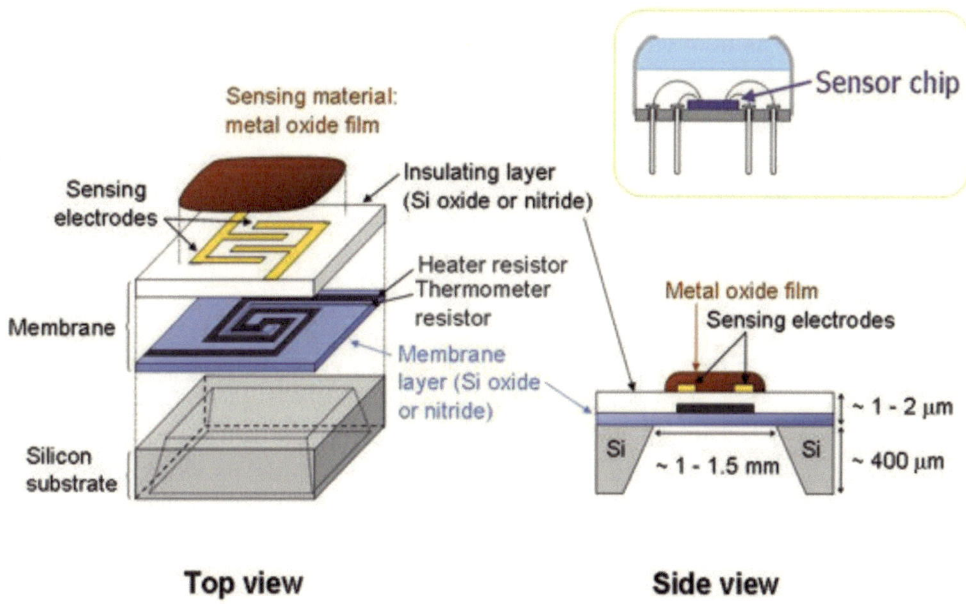

Fig. 4.2 A schematic picture of modern resistive semiconductor gas sensor (Aalto University, Research Group Gas sensors).

This kind of response of electrical conductivity to humidity changes has been extensively investigated for application as humidity sensors. As the response is related to surface adsorption, humidity sensors made using such materials generally exhibit fast response. However, the electronic conductivity can also be influenced by other gaseous species as H_2, CO, to name a few, which can be adsorbed along with H_2O. As a result, there can be significant interference from other species, and such sensors do not exhibit the selectivity required of a practical sensor (Wang and Virkar, 2004).

Gas sensors as shown in Fig.4.2, are mostly n-type semiconductors. They detect reducing gases in air e.g leak detection in gas pipelines, indication of petrol vapour in filling stations or for alcohol tests in exhaled air. The working temperature of sensors are 200-600°C. The grains of the sintered body are covered by adsorbed oxygen. It withdraws electrons from the bulk, forming O^{2-} ions at the surface. Charge carrier concentration in the grain volume decreases, potential barrier is formed at the grain boundaries. The electrical conductance is decreased by oxygen adsorbtion. The molecules of reducing gas interact with adsorbed oxygen lowering potential barrier, increasing conductance of the sensor as follows

$$2CO + O^{2-} = 2CO_2 + e^- \tag{3-7}$$

Numerous reducing gases like hydrogen, methane, CO, ethanol vapor, H_2S can be determined in air (Gründler, 2007).

For humidity sensing, ceramic materials have mainly been used in the form of porous sintered bodies.Thus, controlling porosity and surface activity is of great importance in determining the humidity-sensitive electrical properties of ceramic products (Kim and Gong, 2005).

4.1 Barium titanate as humidity/gas sensor

Ceramic humidity sensors can be classified into the semiconducting and protonic (or ionic) type according to their conduction mechanism. The semiconducting type humidity sensor, operating at elevated temperature (300-400°C), was known to exchange charge carriers (e- and h') between the adsorbed water molecule and ceramic body. While the protonic humidity sensor, usually operating at room temperature, employs protonic (H^+) conduction on the surface of the sample as a sensing mechanism. Both sensors require porous ceramic body for the adsorption of water molecule (Hwang and Choi, 1997).

To make an improved humidity sensor, the microstructure and electrical conductivity of the material should properly be controlled. In general, ceramic humidity sensors are made to be porous and electrically resistive. In this respect, $BaTiO_3$, as a protonic ceramic humidity sensor, has advantage in establishing the microstructure-property relation since the microstructure and conductivity of $BaTiO_3$ can be easily controlled by varying the amount of donor dopant such as La and the processing conditions (Hwang and Choi, 1997).

Ceramic sensors are based on the adsorption of water molecules in the pores, and variations of the structure of the adsorbed layer, as a function of humidity. The sensing mechanism of barium titanate ceramic humidity sensors at room temperature can be explained using ionic conduction model. When few water molecules are available at low humidity, they chemisorb on grain surfaces of the ceramic to form hydroxyl groups as surface charge carriers. When initial water molecules are adsorbed, each water molecule is hydrogen-bonded to two hydroxyls, and the dominant surface charge carriers will be H_3O^+. When still more water is adsorbed, clustering of the water molecules takes place, forming a liquid-like multilayer film of hydrogen-bonded water molecules, where each water molecule is only singly bonded to a hydroxyl group. Since dissociation of H_3O^+ into H_2O and H^+ is energetically favorable in liquid water, the dominant charge carrier in high moisture environment is H+. These processes result in progressively decreasing impedance of barium titanate ceramic when exposed to increasing humidity (Yuk and Troczynski, 2003).

Nanocrystalline $BaTiO_3$ material has better humidity sensing properties than ceramic $BaTiO_3$, such as lower resistance and high sensitivity. However, from the application point of view, one needs to further reduce the resistance of the resistive humidity sensor from 10^6 to 10^3, to minimize the humidity hysteresis and to increase the sensitivity and repeatability(Wang, 2000).

Since water is a polar molecule, the negatively charged oxygen is electrostatically attached to the positively charged Ba^{2+} ions of the sensor material. The initial monolayer is chemisorbed due to the formation of a chemical bond under the influence of high electrostatic field between Ba^{2+} and oxygen of the water layer. This layer, once formed, is not further affected by exposure to humidity. The irreversible reaction for the first layer can be given as

$$Ba^{2+} + H_2O = Ba^{2+} - OH + H^+ \tag{3-8}$$

Once the first layer is formed, subsequent layers of water molecules are physically adsorbed. This physically adsorbed water molecules dissociate because of the high electric fields in the chemisorbed water layer as given below:

$$2H_2O = H_3O^+ + OH \tag{3-9}$$

The charge transport occurs when H_3O^+ ions release a proton to neighbouring water molecules which accepts it while releasing another proton and so on. This is known as Grothuss's chain reaction (Agarwal and Sharma, 2002).

5. Porous ceramics

Porous ceramics find nowadays a large variety of applications in several devices like filters, acoustic absorbers, catalytic components, selective membranes, humidity sensors, among others. There are two main kinds of porous ceramics: reticulate ceramics and foam ceramics. The former consists of interconnected voids surrounded by a web of ceramic; the latter has closed voids within a continuous ceramic matrix. Various chemical routes, besides the conventional solid state reaction route, have been developed for obtaining porous ceramics without the use of additives to avoid unwanted impurities in the final ceramic piece. An attractive alternative route, due to its simplicity and low cost, is the polymeric precursor technique based on the Pechini's patent, that results in sinteractive homogeneous powders. Solid state sintering usually results in polycrystalline bodies with pores in the 100–1000 nm range. Graphite has also been used successfully for increasing pore volume in ceramics. In conventional ceramic sintering, by heating pressed powders, a porous network is usually formed in the spaces between the particles necks, porous stability being achieved by controlling the size of the particles. An easy way of controlling pore sizes during solid state sintering is to control particle sizes of the mixing powders as well as the sintering temperature and time. A previous dilatometric analysis enables one to choose the suitable sintering temperature and time (Cosentino, 2003).

5.1 The definition of porosity and porosity of ceramics

The ceramics are made up of ionic and covalent crystals. However, real crystals are not perfect but contain imperfections that are classified according to their geometry and shape. Vacancy is an example of point defect, dislocation is a line defect, grain boundary is a planar defect (Barsoum 1997). Pores are important features for the ceramics because powder metallurgy methods i.e mixing, milling, shaping and sintering, always result in some amount of porosity. Fig.5.1 indicates atomic mechanisms occurring during sintering. The macroscopic driving force operative during sintering is the reduction of the excess energy associated with surfaces. This can happen by (1) reduction of the total surface area by an increase in the average size of the particles, which leads to coarsening and/or the elimination of solid/vapor interfaces and the creation of grain boundary area, followed by grain growth, which leads to densification (Barsoum 1997).

Pores are three dimensional or volume defects along with cracks, precipitates and particles. Pores generally contain some gas inside them. The reason for this is the employment of an sintering atmosphere during processing. However, it is regarded as a vacuum for simplicity. Pores might be present inside the matrix phase i.e isolated pores and surfaces or they might be located at the grain boundaries and grain junctions.

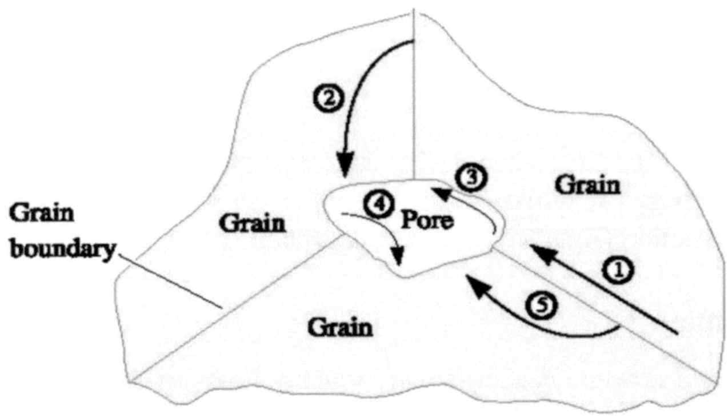

Fig. 5.1 Different diffusion mechanisms involved in sintering. The grain boundary and bulk diffusion (1, 2 and 5) to the neck contribute to densification. Evaporation-condensation (4) and surface diffusion (3) do not contribute to densification (Chiang 1997).

Three quantitites relating to pores are important; the size of the pores, distribution of porosity and total amount of porosity. The shape of the pores is less important. By knowing the constituent phases of the ceramic, theoretical and actual densities can be used to obtain porosity. Mercury intrusion porosimetry can be used to determine pore size and porosity distribution (Carter 2007).

Pores are usually quite deleterious to the strength of ceramics not only because they reduce the cross-sectional area over which the load is applied, but more importantly because they act as stress concentrators (Barsoum 1997). First, they produce stress concentrations. Once the stress reaches a critical level, a crack will form and propagate. Because ceramics possess no plastic-deformation, once a crack is initiated (started), it propagates (grows) until fracture occurs. Second, pores (i.e., their size, shape, and amount) reduce the strength of ceramics because they reduce the cross-sectional areas over which a load can be applied and, consequently, lower the stress that these materials can support (Jacobs 2001).

Porosity can be converted from the density of the ceramics, measured by Archimedes' principle. The density can be calculated from the weight determinations as follows:

$$\rho = \{m_0 / [m_1 - (m_2 - m_w)]\} \times \rho_{water} \quad (5\text{-}1)$$

where m_0 is the dry weight of the samples, m_1 is the weight after dipping into water for 4h. and fluid impregnation and m_2 is the weight while immersed in water. mw is the weight of the wire used to suspend the samples in water, which is also measured in water. All weights are in grams. ρ_{water} is the density of water (in g cm $^{-3}$), which is ρ_{water} = 1.0017-0.0002315T with T is water temperature in °C. The theoretical density of $BaTiO_3$ is considered to be 6.0 g cm^{-3} (Zhang, 2004).

5.2 Different kinds of pore-forming agents for porous BaTiO3 fabrication

PTCR applications

Porous semiconducting $BaTiO_3$ can be prepared by the thermal decomposition of barium titanyl oxalate $BaTiO(C_2O_4)_2.4H_2O$ or by the incorporation of graphite, polyvinylalcohol (PVA), polyvinylbutyral (PVB), borides, silicides, carbides, partially oxidized Ti powders and potato-starch to $BaTiO_3$. These $BaTiO_3$ ceramics indicate large PTCR effects. Oxygen can be adsorbed at the grain boundaries due to the presence of pores. Porous ceramics are more favorable to the formation surface of acceptor states compared to dense ceramics. Porous thermistors show better heat resistance than dense ones and thus can be used as overcurrent protectors in electric circuits. (Kim, 2002, 2003)

For PTCR applications, among these agents polymers, different kinds of starch and metallic additives can be employed. The polymer additives include polyvinylalcohol (PVA), polyvinylbutyral (PVB) and (PEG). The starch can be in the sort of potato or corn starch. Partially oxidized titanium can be another choice to enhance porosity. Carbon addition can also result in the porosity increase. For humidity sensing applications, graphite and PMMA additions can be made to enhance porosity. The effects of different groups of PFAs on the grain size, porosity, PTCR and sensing property of barium titanate ceramics will be mentioned in the following paragraphs. Some of the examples of PFAs are given below, the detailed data on all the PFAs will be presented in the following paragraphs.

5.2.1 The effect of PEG on PTCR properties

Polyethylene glycol (PEG) with amount ranging from 1 to 20 wt.% was added to $BaTiO_3$. The porosity and the grain size as shown in Fig. 5.2, of the n-$BaTiO_3$ ceramic containing 5 wt.% PEG is 10.4 % and 6.3 μm, respectively, while it is 25.2 % and 5.0 μm, respectively for the sample containing 20 wt.% of PEG. The crystalline structure of the n-$BaTiO_3$ ceramics at room temperature was tetragonal and independent on the PEG content. PTCR jump of $BaTiO_3$ ceramics containing 1 and 20 wt.% of PEG was measured to be $2.84x10^5$ and $6.76x10^5$, respectively as shown in Fig.5.3 (Kim, 2004).

5.2.2 The effect of starch based agents on PTCR properties

The potato-starch is another pore-former used in the production of $BaTiO_3$ ceramics. 1-20 wt.% of potato-starch was employed to fabricate porous microstructure. The maximum porosity and minimum grain size was obtained for the sample containing 20 wt.% of potato-starch, which were 37.2% and 3.4 μm, respectively. Pore-former increased porosity and decreased the grain sizes (Kim, 2002).

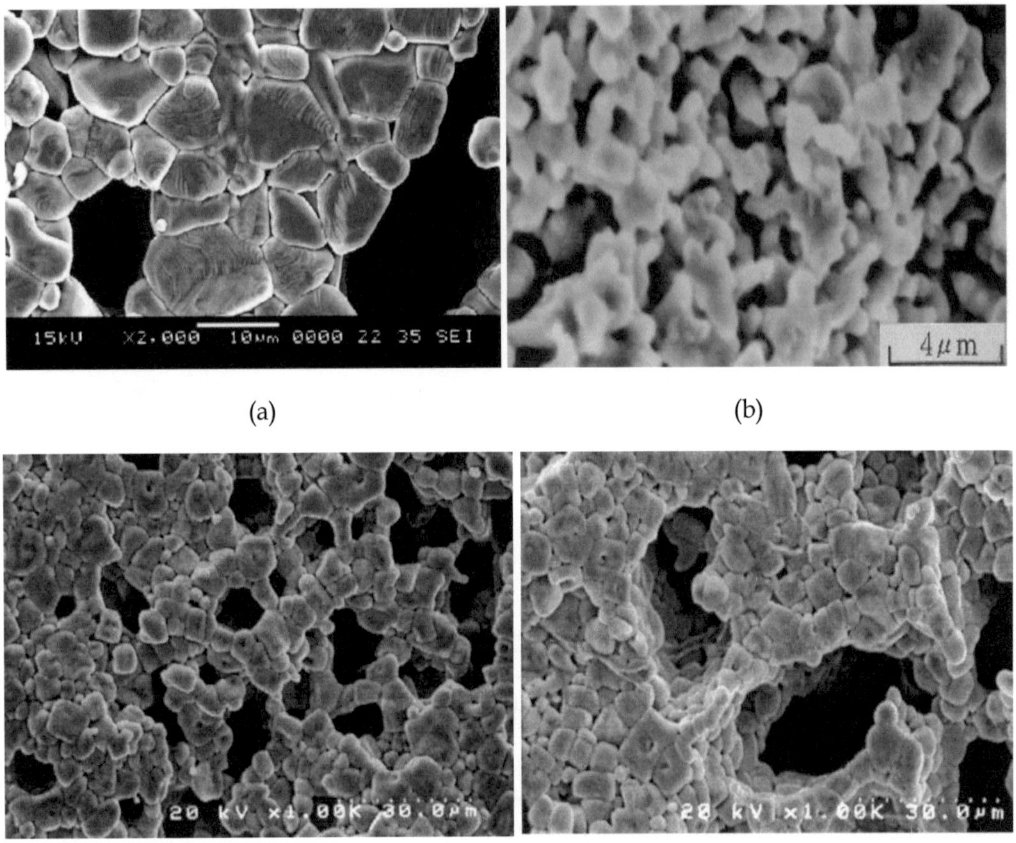

(a) (b)

(c) (d)

Fig. 5.2 Porous microstructures of barium titanate ceramics with the addition of (a) PEG (Kim, 2004) , (b) TiO_2 (Kim et al, , 2002), (c) corn-starch (Kim, 2002) and (d) potato-starch (Kim, 2002).

1-20 wt.% of corn-starch was added into $BaTiO_3$ to produce porosity. The cavities formed due to the burning-out of corn-starch during sintering act as the sites of the pore generations as shown in Fig.5.2. The maximum porosity and minimum grain size, which were 44.0% and 3.1 µm, respectively was measured for the sample containing highest amount of corn-starch. The samples containing corn-starch had a tetragonal crystal structure. The addition of corn-starch into Sb-doped $BaTiO_3$ ceramics results in fine microstructures (3.1-3.8 µm). The porosity and pore size increase and the grain size slightly decreases with corn-starch. The porosity increases from 7.2 to 44% with corn-starch content from 0 to 20wt.%. The grain size changed from 4.8 to 3.1µm. These porous ceramics are advantageous to oxidize grain boundaries and to produce surface acceptor states (Kim, 2002).

The PTCR jump of the Sb-doped $BaTiO_3$ ceramics with corn-starch is observed to be over 10^6 and 1–2 orders of magnitude higher than that of samples without corn-starch as given in Fig.5.3. The donor concentration i.e. Sb and grain size decrease and the electrical barrier height of grain boundaries and the porosity increase with corn-starch content. As a result, room-temperature resistivity increases with pore-forming agent content. The grain

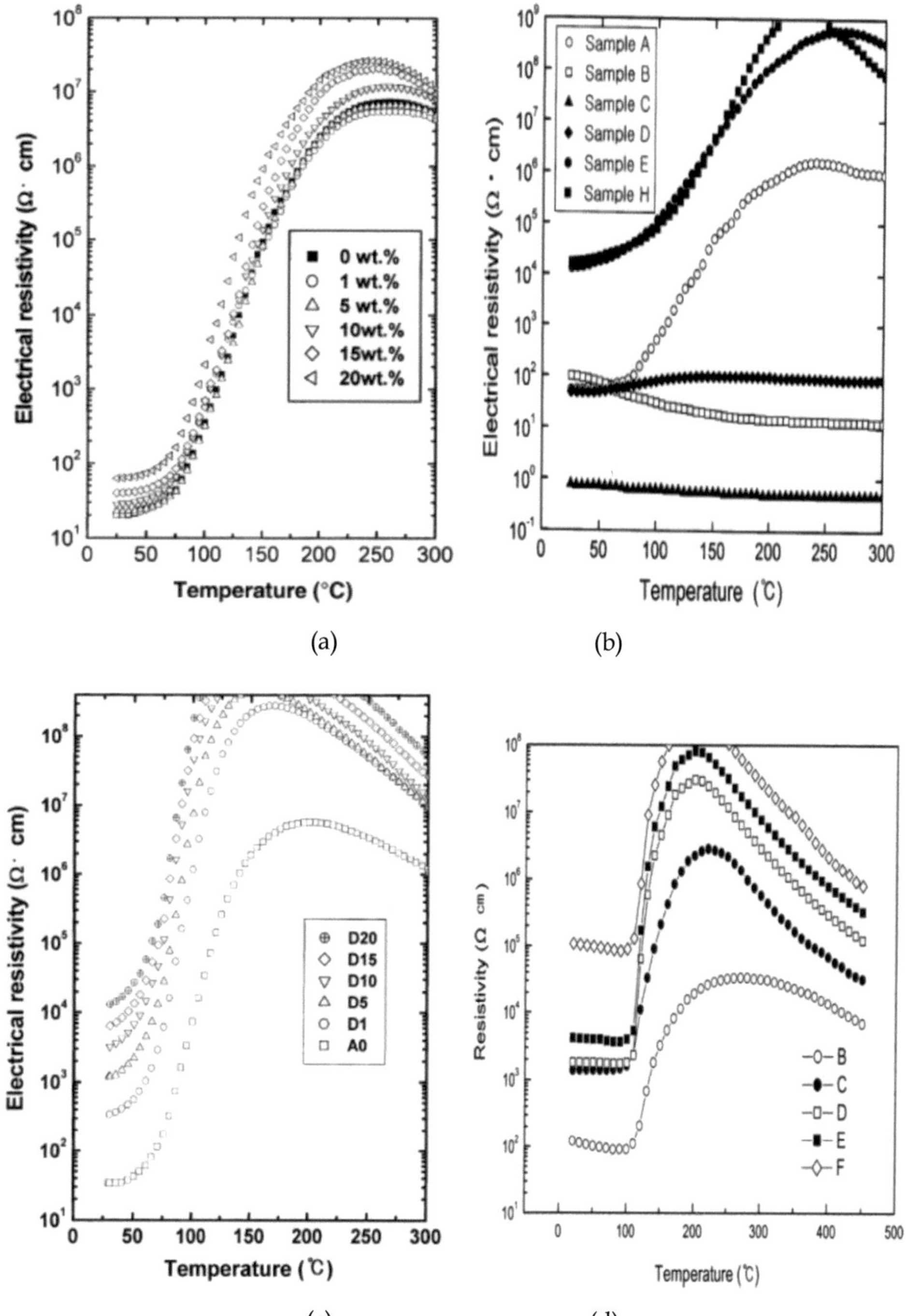

Fig. 5.3 PTCR properties of barium titanate ceramics with the addition of (a) PEG (Kim, 2004), (b) TiO_2 (Kim et al, , 2002), (c) potato-starch (Kim, 2002) and (d) graphite (Park, 2004).

boundary resistivity contributes largely to the total resistivity. The electrical resistivity of grain boundaries significantly depends on the porosity. It can be deduced from this study that, adsorption of chemically adsorbed oxygen atoms at the grain-boundaries increases potential barrier height. Thus electrical resistivity of grain boundaries increases contributing to total resistivity and resulting in a PTCR effect (Kim, 2002).

The electrical resistivities of Sb doped $BaTiO_3$ ceramics were measured during heating under different atmospheres. The high-temperature resistivity under O_2 was higher than that under both air and reducing atmospheres of N_2 and H_2. The low resistivities obtained at high temperatures in reducing atmospheres were due to desorption of chemisorbed oxygen at the grain boundaries, decreasing potential barrier height. Besides, N_2 and H_2 gases diffuse along the grain boundaries and react with chemisorbed oxygen atoms thus chemisorbed oxygen is consumed and conduction electrons are released. This leads to a decrease in potential barrier and as a result, PTCR effects were degraded.

The room-temperature resistivity of porous samples increased with corn-starch content. The room-temperature and high temperature (above Curie temperature) resistivity values were measured to be $1.37x10^2$ and $1.52x10^8$ Ωcm, respectively for the sample containing 1 wt.% of corn-starch. For the rest of the samples, high-temperature resistivities were well above 10 8 ohmcm. Besides, the resistivities of the grain boundaries increased with increasing temperature. The room-temperature and high-temperature resistivities were measured to be $3.52x10^2$ and $2.79x10^8$ Ωcm for the sample containing 1 wt.% of potato-starch. PTCR jump was determined to be $7.93x10^5$ for the same composition. The grain boundary resistivity jump was $1.22x10^3$ for this sample (Kim, 2002).

5.2.3 The effect of graphite agents on PTCR properties

0.3-2.0 mol of C (as graphite powder) was added to Sb doped $BaTiO_3$ to produce porous ceramics. As-sintered samples indicated tetragonal perovskite crystal structure, irrespective of the added graphite content. The porosity of the ceramics increased with graphite addition by an enhanced evolution of CO and CO_2 gases associated with the exothermic reactions. The highest porosity was determined to be 18.3%. The grain growth occurred during sintering due to exothermic reactions and maximum grain size obtained was 160.7 µm. However, as the graphite content increased, the grain sizes decreased. This is because the existing graphites act as barriers to inhibit the grain growth during sintering, reducing the grain size (Park, 2004).

The addition of graphite into $BaTiO_3$ substantially improved PTCR characteristics. For the sample containing highest graphite, the maximum resistivity above Curie temperature was measured to be above 10^8 Ωcm and PTCR jump was above $1.2x10^3$. The maximum room-temperature resistivity was measured for this sample, which was $1.35x10^4$ Ωcm and PTCR jump was $5.86x10^3$. The magnitude of the PTCR effect of samples containing graphite increased by 1-2 orders compared to the sample without porosifier as shown in Fig. 5.3. For the porous samples, fast response to overcurrent (below 20s) was achieved (Park, 2004).

5.2.4 The effect of metallic agents on PTCR properties

Partially oxidized Ti powders were used as pore-former in a study of Kim, Cho and Park. Ti powders were heated at 600°C for 1h in flowing oxygen gas and were covered with TiO_2.

TiO_2(Ti) powders with various contents (0-9 vol%) were added to (Ba,Sr)TiO_3 in vacuum. Vacuum-sintered ceramics were exposed to paste-baking treatment in air in order to make oxygen adsorption possible at the grain boundaries. The porosity increased and grain size decreased by the addition of TiO_2(Ti) powders. The interfaces between TiO_2(Ti) powder of a large size and $BaTiO_3$ powder of a small size provide the sites of the pore generations, resulting in porosity increase. 46.0% and 1.0μm are the porosity and grain size values, respectively for the sample containing 9 vol% of TiO_2(Ti). Small grain size may be due to a decrease in the driving force for the grain boundary movement with TiO_2(Ti) addition. TiO_2(Ti) addition and vacuum sintering (low oxygen content and small grain size) transformed the tetragonal crystal structure of $BaTiO_3$ to cubic. (Ba,Sr)TiO_3 ceramics containing 5 vol% of TiO_2(Ti) showed excellent PTCR characteristic, i.e. low room-temperature resistivity ($1.8x10^{12}$ Ωcm) and the high ratio of maximum resistivity to minimum resistivity ($1.5x10^{16}$) as shown in Fig. 5.3 (Kim et al, , 2002).

Humidity/Gas sensing Applications

5.2.5 The effect of graphite agents on humidity sensing

The effect of graphite addition on grain size, porosity and humidity sensing properties was investigated. Four different graphite containing compositions (C-1/C-4) were sintered at 1200-1500°C for 2-6h.

The graphite addition increased the grain size of the donor doped samples due to heat generated by the exothermic reactions of carbon combustion. However, the grain size decreased by further addition of graphite powder, because of the graphite segregation at the grain boundaries (Ertug et al., 2005). This result is in good agreement with (Park, 2004)'s study on PTCR properties.

The addition of graphite resulted in a porous microstructure as shown in Fig.5.4. The maximum grain size was measured for the sample (BTLC-1) sintered at 1500°C for 6h. and it was 15.7 μm. Upon sintering at 1500°C for 6h., the grain size of BTLC-4 composition reached 13.3 μm. The grain size decreased drastically when the sintering temperature was decreased from 1500°C to 1200°C for 2-6h. For the sample sintered at 1200°C for 6h., the grain size was measured to be 5.7 and 5 μm for BTLC-1 and BTLC-4 compositions, respectively (Ertug et al., 2005).

The porosity percentage of the graphite containing compositions changed dramatically when the sintering temperature was increased from 1200 to 1500°C. The porosity change was slight between 1200-1400°C however after sintering at 1500°C, most of the porosity in the microstructure was eliminated and dense samples were obtained (Ertug et al., 2005).

The minimum porosity percentage was measured for the sample (BTLC-1) sintered at 1500°C for 6h. and it was 12 %. When graphite addition was increased from 3.5 to 6.5%, the porosity percentage changed from 59 to 64.9 %upon sintering at 1200°C for 6h (Ertug et al., 2005).

The humidity sensing properties of graphite containing compositions were given in Fig.5.5 and 5.6. The transfer function of the humidity sensitive samples, which were sintered at 1200°C was of exponential type. At the low humidity percentages, the electrical resistance decreased with relative humidity linearly. When the relative humidity increased beyond the

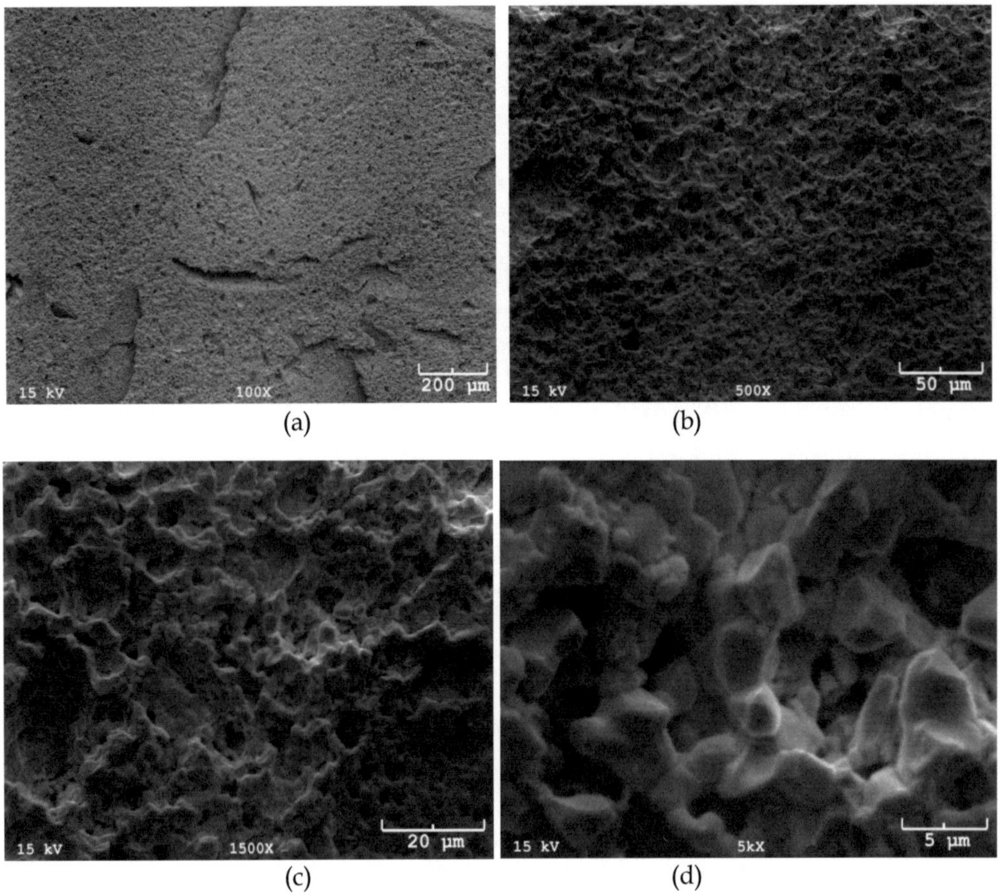

Fig. 5.4 The porous microstructures of graphite containing barium titanate samples (Ertug, 2008).

critical value, the electrical resistance-relative humidity curve became exponential. The minimum electrical resistance was measured for the samples which contain higher porosity percentages. When the content of the graphite was increased, the transfer function of the humidity sensitive sample did not change, i.e exponentail curve. However, the electrical resistance values indicated a slight decrease due to higher porosity percentage (Ertug, 2008).

When the samples were sintered at 1500°C, porosity percentages decreased by a great amount. As shown in Fig.5.6, as-sintered samples did not exhibit a change in the electrical resistance with relative humidity (Ertug, 2008).

5.2.6 The effect of PMMA agent on humidity sensing

The three different barium titanate compositions containing PMMA as pore forming agent were investigated. The samples were sintered at 1200-1500°C for 2-6h. The grain size of the samples decreased with PMMA content. The maximum grain size was measured for the sample sintered at 1500°C for 6h. and it was 14.87 µm. The maximum porosity was determined for the sample containing maximum PMMA amount, which was 53.5 %(Ertug, 2006).

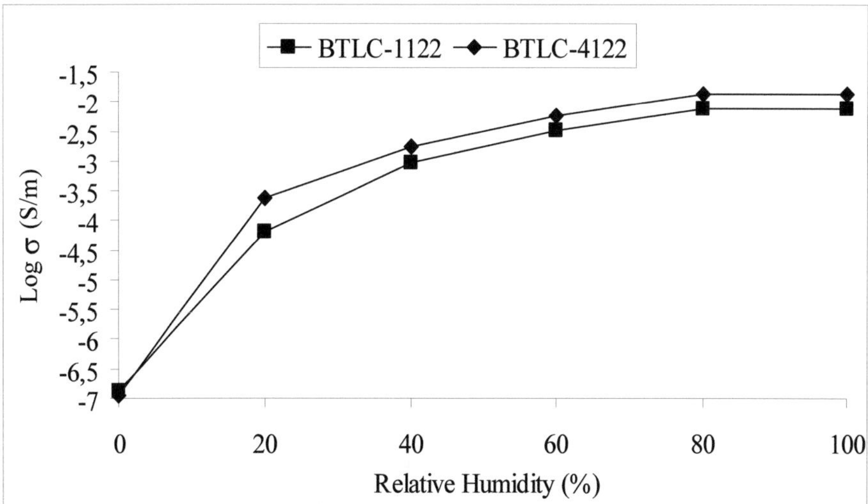

Fig. 5.5 The humidity sensing property of graphite containing samples sintered at 1200°C (Ertug, 2008).

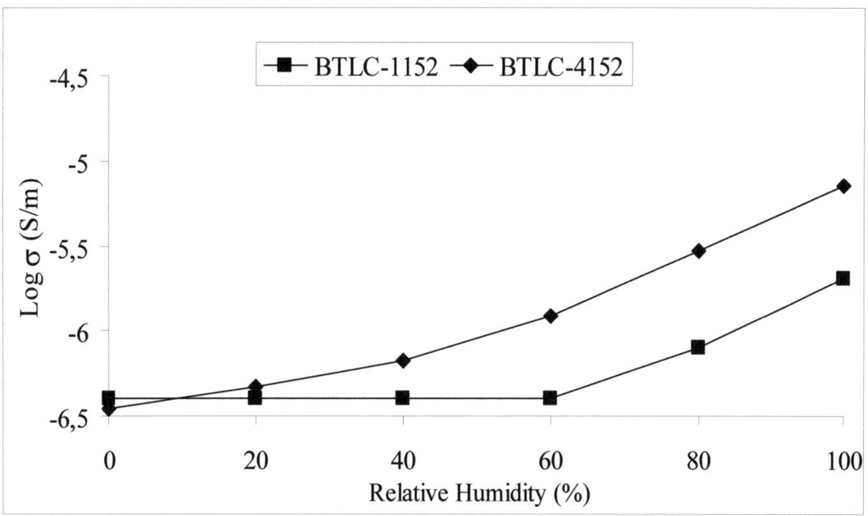

Fig. 5.6 The humidity sensing property of graphite containing samples sintered at 1500°C (Ertug, 2008).

The transfer functions of the barium titanate samples changed with sintering temperature. When the samples were sintered at 1200°C, the transfer function was exponential. However, when the sintering temperature was increased to 1500°C, the transfer function of the composition containing lower PMMA, included a dead band region where at low relative humidity, the electrical resistance remained constant with humidity. The dead band region was valid up to 40% of relative humidity. As shown in Fig.5.7., the composition which contained higher amount of PMMA, did not show a certain dead band region (Ertug, 2008).

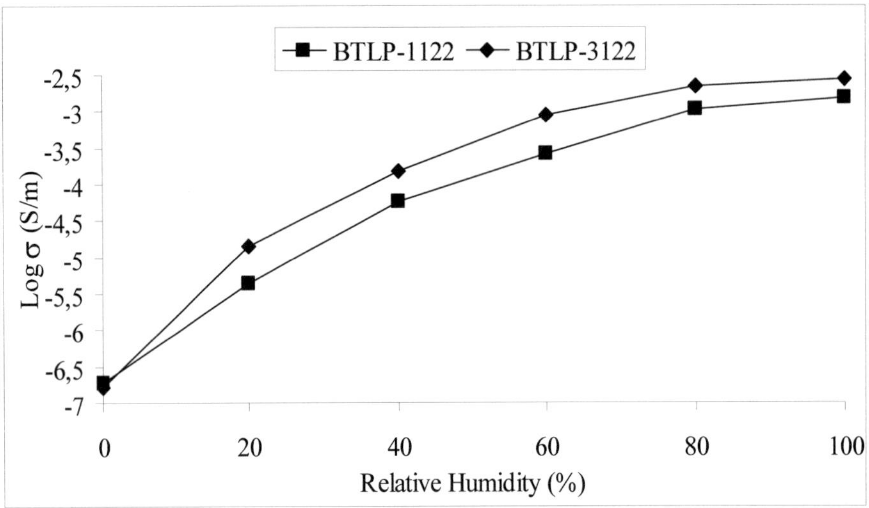

Fig. 5.7 The humidity sensing property of PMMA containing samples sintered at 1200°C (Ertug, 2008).

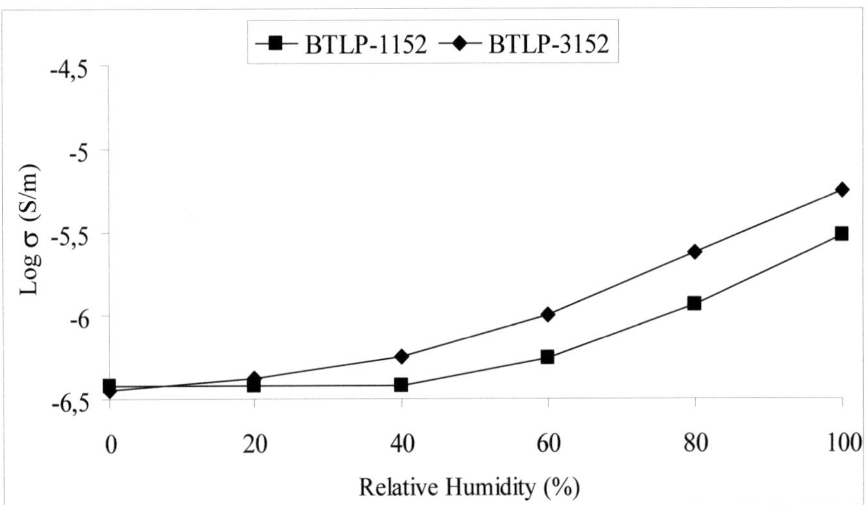

Fig. 5.8 The humidity sensing property of PMMA containing samples sintered at 1500°C (Ertug, 2008).

However, the electrical resistance did not change very much with relative humidity up to 40% of relative humidity. A clear drop in the electrical resistance was observed only after 60% of relative humidity (Ertug, 2008).

6. Conclusion

n-type semiconducting barium titanate ceramics have applications as thermistors and sensors. For barium titanate in order to exhibit PTCR and gas (or humidity) sensing property, first donor doping by trivalent or pentavalent cations is necessary. The improvement of porosity is another requirement for PTCR and sensor applications. Porous barium titanate ceramics are generally fabricated using different kinds of pore-forming agents (PFAs). The fabrication process is carried out using powder metallurgy methods namely mixing with PFAs, shaping and sintering. PFAs, which are used for porous barium titanate fabrication, can be grouped into several categories depending on the type of the pore-former. Polymeric agents, metallic agents, starch-based agents and carbon-based agents. An additional binder burn out step is employed for some of the pore-formers such as PEG and direct sintering is done for other pore-formers such as starch and graphite. The pore-forming mechanisms of several PFAs are different from each other. When oxidized Ti powders are used as PFAs, the difference in the powder size of TiO_2 (18μm) and $BaTiO_3$ (1m) powders lead to pore generation. However, gas evolution during the burning of carbon results in porosity when graphite PFA is used. When polymeric agents are used to produce porosity, organic components are pyrolized upon binder removal process. The porosity improvement by PFAs lead to humidity or reducing gas sensitivity of barium titanate based ceramics. Besides, the oxidation of the grain boundaries are enhanced by porosity increase which in turn improves PTCR property. Brand new PFAs such as several polymeric agents, can be used to produce porous barium titanate or known to be effective PFAs can be applied to it in order to produce more sensitive sensors or to obtain more pronounced PTCR effect in $BaTiO_3$. The effect of PFAs on the porosity (and pore size) and grain size of barium titanate ceramics and also the different pore-forming mechanisms of PFAs are currently open to further investigation since this particular area is based on trial and error method.

7. Acknowledgment

Dedicated to my beloved sister, Deniz for supporting me through my academic life.

I wish to thank my Ph.D. thesis advisor Prof.Dr. A. Okan Addemir at the Istanbul Technical University, who enabled the beginning and the completion of the work on porous barium titanate based humidity sensors.

I would like to express my gratitude to Assistant Prof. Dr. Tahsin Boyraz at Cumhuriyet University, who gave me the idea of working on and writing a book chapter on powder metallurgy. I truly appreciated all the time and advice he gave me throughout my time at ITU.

I also wish to thank my collegue at ITU, Gülten Sadullahoğlu, Ph.D. for her constant support in my studies.

8. References

Agarwal, S. and Sharma, G.L. (2002). Humidity sensing Properties Of (Ba, Sr)TiO3 thin films grown by hydrothermal-electrochemical method, *Sensors and Actuators B*, 85, 205-211.

Barsoum, M.W. (1997). *Fundamentals Of Ceramics,* The McGraw-Hill Companies, Inc., pp.149-411.

Brook, R.J. (1991). *Concise Encyclopedia of Advanced Ceramic Materials,* Pergamon Press.

Carter, C.B. and Norton, M.G. (2007). *Ceramic Materials Science and Engineering,* Springer Science + Business Media LLC, pp. 277-279.

Chen, Y.C. (2007). Annealing Effects Of Semiconducting Barium-Titanate Thermistor, *Journal of Marine Science and Technology,* Vol. 15, No. 4, pp. 307-314 .

Chiang, Y.M., Birnie, D. and Kingery, W.D. (1997). *Physical Ceramics: Principles for Ceramic Science and Engineering,* John Wiley & Sons, Inc.

Cosentino, I.C., Muccillo, E.N.S. Muccillo, R. (2003). Development of zirconia-titania porous ceramics for humidity sensors, *Sensors and Actuators B*, 96, pp. 677-683.

Desu, S.B. (1990). Interfacial Segregation in Perovskites: III, Microstructure and Electrical Properties. *J. Am.Ceram.Soc.,* 73, pp. 3407-15.

Ertug, B., Boyraz, T., Addemir, O., (2005). Effect of graphite content on the porosity and microstructural characterization in La doped $BaTiO_3$" ,*IX Conference & Exhibition of the European Ceramic Society*, 19 - 23 June, Portoroz, Slovenia.

Ertug, B., (2008). The Effect of Pore-Forming Agents on the Electrical Properties of Barium Titanate Based Ceramics Under Humid Environment, Istanbul Technical University, Institute of Science and Technology.

Ertug, B., Boyraz, T., Addemir, O., (2006). PMMA Katkısının $BaTiO_3$ Esaslı Poroz Seramiklerin Porozite ve Mikroyapısal Özellikleri Üzerine Etkisinin İncelenmesi, *Materials 2006, 11th International Materials Symposium,* 19-21 April, Denizli-Turkey.

Fine, G.F., Cavanagh, L.M., Afonja, A. and Binions, R. (2010). Metal Oxide Semi-Conductor Gas Sensors in Environmental Monitoring, *Sensors,* 10, pp. 5469-5502.

Gründler P. (2007). *Chemical Sensors,* An Introduction for Scientists and Engineers, Springer.

Hutagalung, S.D. (2011). Technical Ceramics; Ferroelectric Ceramics Lecture Notes, School of Materials and Mineral Resources Engineering, Universiti Sains Malaysia.

Hwang, T.J. and Choi, G.M. (1997). Electrical characterization of pororus BaTiO3 using impedance spectroscopy in humid condition, *Sensors and Actuators B,* 40, pp. 187-191.

Jacobs, J.A. and Kilduff, T.F. (2001). *Engineering Materials Technology,* Prentice-Hall, Inc. pp. 430-431.

Jiang, S., Zhou, D., Gong, S. and Guan, X. (2003). Effect of heat-treatment under oxygen atmosphere on the electrical properties of BaTiO3 thermistor, *Microelectronic Engineering,* 66, pp. 896-903.

Kim, J.G., Cho, W.S. and Park, K. (2000). PTCR characteristics in porous (Ba,Sr)TiO3 ceramics produced by adding partially oxidized Ti powders, *Materials Science and Engineering B,* 77, pp. 255-260.

Kim, J.G. (2002) Effect of atmosphere on the PTCR characteristics of porous, Sb-doped BaTiO3 ceramics produced by adding potato starch. *Journal of Materials Science Letters,* 21, pp. 1645-1647.

Kim, J.G. (2002). Synthesis of porous (Ba,Sr)TiO3 ceramics and PTCR characteristics, *Materials Chemistry and Physics,* 78, pp. 154–159.

Kim, J.G., Cho, W.S., Park, K. (2002). Effect of reoxidation on the PTCR characteristics of porous (Ba,Sr)TiO3, *Materials Science and Engineering B,* 94, pp. 149–154.

Kim, J.G. (2003). Synthesis of porous (Ba,Sr)TiO3 ceramics and PTCR characteristics, *Materials Chemistry and Physics,* 78, pp. 154-159.

Kim, J.G., Tai, W.P., Kwon, Y.J., Lee, K.J. , Cho, W.S. , Cho, N.H., Whang, C.M. and Yoo, Y.C. (2004). Effects of reducing and oxidizing atmospheres on the PTCR characteristics of porous n-$BaTiO_3$ ceramics by adding polyethylene glycol, *Journal of Materials Science: Materials in Electronics,* Volume 15, Number 12, pp. 807-811.

Kim, J.G., Tai, W.P., Lee, K.J., Cho, W.S. (2004). Effect of polyethylene glycol on the microstructure and PTCR characteristics of *n*-BaTiO3 ceramics, *Ceramics International, 30,* pp. 2223-2227.

Kim, Y.S., Ha, S.C., Kim, K., Yang, H., Choi, S.Y. and Kim, Y.T., Park, J.T., Lee, C.H., Choi, J., Paek, J. and Lee, K. (2005). Room-temperature semiconductor gas sensor based on nonstoichiometric tungsten oxide nanorod film, *Applied Physics Letters,* 86, pp. 213105.

Kim, D.U., Gong, M.S. (2005). Thick films of copper-titanate resistive humidity sensor, *Sensors and Actuators B: Chemical,* Volume 110, Issue 2, pp. 321-326.

Morrison, F.D., Coats, A.M., Sinclair, D.C., West, A.R. (2001). Charge Compensation Mechanisms in La-doped BaTiO3. *Journal of Electroceramics,* 6, pp. 219-232.

Park, K., Kim, J.G. and Cho, W.S. (2002) Effect of partially oxidized Ti powders on the microstructure and PTCR characteristics of (Ba,Sr)TiO_3, *Journal of Materials Science: Materials in Electronics,* Volume 13, Number 1, pp. 13-19.

Park, K. (2004). Characteristics of porous BaTiO3-based PTC thermistors fabricated by adding graphite powders, *Materials Science and Engineering B,* 107, pp. 19-26.

Park, K. (2004). Characteristics of porous BaTiO3-based PTC thermistors fabricated by adding graphite powders, *Materials Science and Engineering B,* 107, pp. 19-26.

Pokhrel, S., Jeyaraj, B., Nagaraja, K.S. (2003) . Humidity-sensing properties of ZnCr2O4–ZnO composites, *Materials Letters,* 57, pp. 3543–3548.

Qi, J., Gui, Z., Wang, Y., Zhu, Q., Wu, Y. and Li, L. (2002). The PTCR effect in BaTiO3 ceramics modified by donor dopant, *Ceramics International,* 28, pp. 141-143.

Smyth, D.M. (2000). The effects of dopants on the properties of metal oxides. *Solid State Ionics,* 129, pp. 5-12.

Steele, B.C.H. (1991). *Electronic Ceramics,* Elsevier Applied Science, Chapman and Hall, New York.

Tsur Y., Dunbar T.D., Randall C.A. (2001). Crystal and Defect Chemistry of Rare Earth Cations in BaTiO3. *Journal of Electroceramics,* 7, pp. 25-34.

Vijatovic, M.M., Bobic, J.D., Stojanovic, B.D. (2008). History and Challenges of Barium Titanate: Part II. *Science of Sintering,* 40, pp. 235-244.

Wagiran, R., Wan Zaki, W.S., Mohd Noor, S.B., Shaari, A.H. and Ahmad, I. (2005). Characterization Of Screen Printed BaTiO3 Thick Film Humidity Sensor, *International Journal of Engineering and Technology,* Vol. 2, No. 1, pp. 22-26.

Wang, W. and Virkar, A.V. (2004). A conductimetric humidity sensor based on proton conducting perovskite oxides, *Sensors and Actuators B,* 98, pp. 282-290.

Wang, J., Xu , B., Liu, G., Zhang, J., Zhang, T. (2000). Improvement of nanocrystalline BaTiO humidity sensing properties, *Sensors and Actuators B,* 66, pp. 159-160.

Wang, H.L. (2002) Structure and Dielectric Properties of Perovskite-Barium Titanate (BaTiO3), Submitted in Partial Fulfillment of Course Requirement for MatE 115, San Jose State University.

Wegmann, M., Brönnimann, R., Clemens F. and Graule, T. (2007). Barium titanate-based PTCR thermistor fibers: Processing and properties, *Sensors and Actuators A: Physical,* Volume 135, Issue 2, pp. 394-404.

Yuk, J., Troczynski, T. (2003). Sol–gel BaTiO3 thin film for humidity sensors, *Sensors and Actuators B,* 94, pp. 290-293.

Zhang, D., Zhou, D., Jiang, S., Wang, X. and Gong, S. (2004). Effects of porosity on the electrical characteristics of current-limiting BaTiO3-based positive-temperature-coefficient (PTC) ceramic thermistors coated with electroless nickel-phosphorus electrode, *Sensors and Actuators A,* 112, pp. 94-100.

Zhou, L. (2000). Processing Effects on Core-Shell Grain Formation in ZrO2 Modified BaTiO3 Ceramics, Ph.D Thesis, University of Cincinati.

Springer Science+Business Media, Available from: http://www.springerimages.com/

Electronic Circuits and Projects Forum, Available from: http://www.electro-tech-online.com/

Aalto University, Department of Micro and Nanosciences, Research Groups Gas sensors, Available from: http://nano.tkk.fi/en/research_groups/electron_physics/research/gas_sensors/

5

Hybrid Gas Atomization for Powder Production

Udo Fritsching and Volker Uhlenwinkel
University of Bremen
Germany

1. Introduction

Atomization of a molten liquid melt is a versatile method for powder production. Main operational parameters and physical conditions for a suitable atomization process for different applications in powder production are:

- melt type: surface tension, viscosity and temperature range (solidification temperature),
- process type: aimed throughput, energy efficiency,
- product type: particle size distribution (mean particle size and width of distribution).

Most atomization processes e.g. for metal powder production are gas atomization processes using twin fluid atomizers. For these applications, besides the use of standard atomizer types, relevant developments are ongoing to derive advanced atomizer concepts for improved powder products and energy and resources effective atomization processes.

Main aims of atomizer developments are:

- minimization of (mean) particle sizes,
- narrowing of the particle size distribution width,
- effective use of resources (gas and feed stock),
- technical production of complex melt systems for powder applications.

In this contribution the development of gas atomization units is described, which utilize characteristic flow effects of a molten metal stream in relation to the flow of compressed gases for efficient fragmentation of liquids and melts. The interaction of gas and liquid melt in the twin-fluid atomization process is used for understanding and optimization of melt fragmentation processes.

2. Atomization of liquids and melts

Atomization of liquids (e.g., pure liquids, solutions, suspensions and emulsions, or melts) is a classic process engineering unit operation. The process of liquid atomization has applications in numerous industrial branches, for example, in chemical, mechanical, aerospace, and civil engineering as well as in material science and technology and metallurgy, food processing, pharmaceuticals, agriculture and forestry, environmental protection, medicine, and others.

Within atomization, the bulk fluid (continuous liquid phase) is transformed into a spray system (dispersed phase: droplets). The disintegration process itself is caused either by intrinsic (e.g., potential) or extrinsic (e.g., kinetic) energy, where the liquid, which is typically fed into the process in the form of a liquid jet or sheet, is atomized either due to the kinetic energy contained in the liquid itself, by the interaction of the liquid sheet or jet with a (high-velocity) gas, or by means of mechanical energy delivered externally, e.g., by rotating devices.

The main purpose of technical atomization processes is the generation of a significantly increased gas-liquid interface in the dispersed multiphase system. All transfer processes across the gas–liquid phase boundary directly depend on the driving potential difference of the exchange property (heat, mass, or momentum) and the size of the exchange surface. This gas–liquid contact area in a dispersed spray system is correlated with the integral sum of the surfaces of all individual droplets within the spray. An increase in the relative interphase area in a dispersed system intensifies momentum, heat, and mass-transfer processes between the gas and liquid. The total flux of exchange within spray systems is thereby increased by several orders of magnitude.

Within the atomization process typically a liquid stream (jet or sheet) is disintegrated into a large number of droplets of various sizes. As all of the solid state materials as minerals, ceramics, glass and metals may be converted into the liquid state by melting, powders from all of these materials may be produced via atomization of the liquid melt and consecutive solidification of the droplets in the spray to achieve a powder.

The area of melt atomization differs in several ways and properties from a conventional atomization process (lets say from water or fuel atomization) due to the physical melt conditions as:

- high temperatures above the relevant melting point
- comparably high surface tensions (metal melts)
- comparably high viscosity (glass or ceramic melts)

The general physics, devices and application of (conventional) atomization and spray processes have been reviewed and intensively published as e.g. in (Lefebvre, 1989; Bayvel & Orzechowski, 1993; Fritsching; 2006; Ashgriz, 2011).

2.1 Gas atomization of liquid melts

Several principle atomisation mechanisms and devices exist for disintegration of molten metals. An overview on molten metal atomisation techniques and devices is given e.g. by (Lawley, 1992; Bauckhage, 1992; Yule & Dunkley, 1994, and Nasr et al., 2002). In the area of metal powder production by atomisation of molten metals typically twin-fluid atomisation by means of inert gases is used. Main reasons for using this specific atomisation technique are:

- the possibility of high throughputs and disintegration of high mass flow rates
- the greater amount of heat transfer between gas and particles for rapid partially cooling the particles
- the direct delivery of kinetic energy to accelerate the particles towards the substrate/deposit for compaction
- the minimization of oxidation risks of the atomised materials within the spray process by use of inert gases.

A common characteristic of the used variations of twin-fluid atomizers for molten metal atomisation is the vertical exit of the melt jet from the tundish via the (in most cases cylindrical) melt nozzle in the direction of gravity. Also in most cases the central melt jet stream is surrounded by a gas flow from a single (slit) jet configuration or a set of discrete gas jets, which are flowing in parallel direction to the melt flow or within an inclination angle of attack towards the melt stream. The coaxial atomizer gas usually exits the atomizer at high pressures with high kinetic energy.

Two main configurations and types of twin-fluid atomizers need to be distinguished within molten metal atomisation. The first kind is called the confined or close-coupled atomizer and the second kind is called the free-fall atomizer. Both concepts are illustrated in Figure 1.

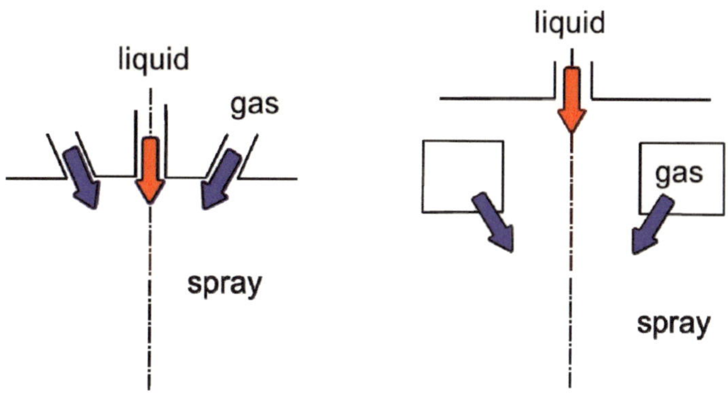

Fig. 1. Gas atomization principles of melts: Close coupled atomizer (left) and free-fall atomizer (right)

The gas flow in the close-coupled atomizer immediately covers the exiting melt jet. Within the confined atomizer the distance between gas exit and melt stream is much smaller than in the free-fall arrangement, where the melt jet moves a certain distance in the direction of gravity before the gas flow impinges onto the central melt jet. The close-coupled configuration generally tends to yield higher atomisation efficiencies (in terms of smaller particles at identical energy consumption) due to the lower distance between the gas and melt exits. But the confined atomizer type is more susceptible for freezing problems of the melt at the nozzle tip. This effect is due to the extensive cooling of the melt by the expanding gas flow, which exits in the close-coupled type nearby to the melt stream. During isentropic gas expansion the atomisation gas temperature is lowered (sometimes well below 0°C). Because of the close spatial coupling between gas and melt flow fields, this contributes to a rapid cooling of the melt at the tip of the melt nozzle. The freezing problem is relevant especially for spray forming applications. In this process, in all technical applications a discontinuous batch operation is performed (e.g. due to the batchwise melt preparation or the limited preform extend to be spray formed). The operational times of spray forming processes are ranging from several minutes up to approximately one hour. The thermal related freezing problem is most important in the initial phase of the process directly when the melt stream exits the nozzle for the first time. At that point the nozzle tip is still cool and needs to be heated first e.g. by the hot melt flow. This heating process lasts a certain time. Therefore, thermal related freezing problems are often to be observed in the first few seconds of a melt atomisation process.

In addition to the thermal related freezing problem within the melt nozzle, also chemical or metallurgical related problems in melt delivery systems are found frequently. Not all of these problems are solved yet in melt atomisation applications. A range of problems arises from the possible change of material composition of the melt, or that of the tundish or nozzle material, due to possible melt/tundish reactions or melt segregational effects from diffusion. This reaction process kinetics is somewhat slower than the thermal freezing process kinetics mentioned before and may contribute to operational problems at a later temporal state of process operation.

Free-fall atomizers are much less problematic than close-coupled atomizers in terms of thermal freezing processes, as the melt jet stream and the gas stream are well separated at the exit of the melt from the delivery system (tundish exit). Therefore, the cooling of the melt due to the cold gas occurs at a later position than within close-coupled atomizers. The free-fall atomizer obeys as an additional advantage in the frame of spray forming the possibility of controlled mechanical or pneumatic scanning and therefore oscillating the gas atomizer with respect to one axis. Also concepts of spatial/temporal distribution of segments of gas jets within the nozzle can be used for atomisation. By doing so, the free-fall atomizer gives the operator an additional degree of freedom with respect to the control and regulation of the mass flux distribution of droplets in the spray. This most important physical property of the spray within other atomizer nozzle systems can (within a running process) only be influenced by changing the atomizer gas pressure. By controlled scanning of the nozzle, the mass flux can also be distributed over a certain area (necessary e.g. for flat product spray forming).

3. Free-fall atomizer design for melts

The design of a conventional free-fall atomizer is simple, robust and very reliable. Thus the free-fall atomizer concept has several applications in liquid processing. Free-fall atomizers are applied e.g. in spray drying facilities, in powder production devices, or in spray forming processes (Yule & Dunkley, 1994; Fritsching, 2008). Figure 1 shows a sketch of a conventional external mixing free-fall atomizer. The design is based on the combination of two gas nozzle systems, namely a primary and a secondary gas nozzle (Fritsching, & Bauckhage, 1992). The secondary nozzle is the main atomization unit. The concentric gas jets from the secondary nozzle impinge onto the central liquid jet that is disintegrated due to instabilities from the shearing action of the secondary gas flow and its relative velocity (Markus et al., 2002). The disintegration process of the liquid jet in the atomization zone is located underneath the secondary gas nozzle (Lohner et al., 2003; Heck, 1998). Due to this separation of the atomization area and the atomizer body, the free-fall atomizer is mainly used for atomization of viscous or high temperature liquids and melts.

Due to the inclination of the secondary gas flow, underneath the secondary gas nozzle a recirculation gas flow area may occur depending on the atomizer design and operation conditions. If the intensity of the recirculation flow is reasonably high, liquid ligaments or droplets are transported back to the atomizer body. Here, these sticking fragments may clog the gas and/or liquid orifices and negatively influence or even stop the atomization process. In common free-fall atomizer designs the recirculation gas flow is suppressed by the use of a primary gas nozzle. The primary gas flow is coflowing to the liquid jet thereby guiding the liquid into the atomization area without recirculation.

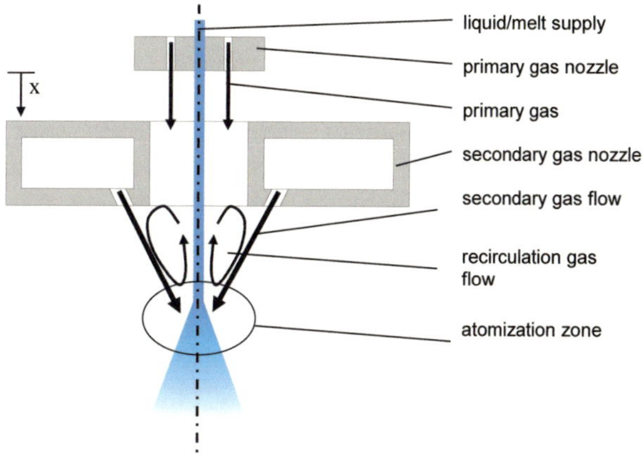

Fig. 2. Main components and areas of a conventional free-fall atomizer (From Czisch & Fritsching, 2008a, with permission)

During operation of the free-fall atomizer, the gas mass flow rates of both gas nozzle systems are controlled by the primary and secondary gas pressures respectively (pressure levels are given as of overpressure) (Czisch & Fritsching, 2004). Problems may occur during atomizer operation if the pressure ratio between the primary and the secondary nozzle is improperly adjusted and an intense recirculation flow is generated. The applicable maximum gas pressure of the primary gas nozzle is limited by initial disturbances that may be generated on the liquid jet before reaching the atomization area. Therefore, the secondary gas pressure is limited also because the secondary gas mass flow rate determines the necessary primary gas mass flow rate for prevention of recirculation. This coupling of the two nozzle systems limits the applicability of the free-fall atomizer (Fritsching, 2004; Heck et al., 2000; Bergmann et al., 2001).

To improve the operating conditions of the free-fall atomizer it is intended to increase the gas pressures while the recirculation flow is still suppressed. In this case the recirculation gas flow underneath the secondary gas nozzle must be prevented in a different way. It is proposed here that a cylindrical ring device installed into the gas flow of the secondary gas nozzle can improve the atomization performance. The applied device utilizes the Coanda effect (Wille & Fernholz, 1965) and the injector principle to influence the local atomization gas direction. Within the confined flow device (called here the Coanda-ring) an under-pressure zone is generated which deflects the gas flow downwards and increases the pressure difference between the inner zone and the ambient gas pressure. This downward reflection of the gas stream effectively increases the gas entrainment from outside the main flow and therefore increases the total gas mass flow rate. This effect increase becomes important because the liquid jet interaction within the secondary nozzle is increased and the recirculation gas flow is suppressed.

3.1 Numerical simulation of free-fall atomizer gas flow

The improvement of the atomizer design and operating conditions is based on numerical investigations of the gas flow field of a conventional free-fall atomizer (primary and secondary gas nozzle; see figure 2). Nitrogen is used througout this study as atomizer gas. By applying a axisymmetric two-dimensional flow simulation it is assumed that the gas exit configuration is that of a circular slit nozzle (Lohner, 2002; Lohner et al., 2005). Compressibility (assuming ideal gas flow) as well as turbulence effects (using the conventional k-eps turbulence model) are taken into account. The gas condition at the nozzle exit is assumed as perfectly expanded (supersonic). Pressure levels are given as overpressure (pressure level exceeding the ambient pressure). The orifice areas of primary and secondary gas are kept constant. Therefore, the gas mass flow rates of the primary and the secondary gas nozzle are varied by changing the stagnation gas pressures for primary and secondary gas, respectively.

In Figure 3 the influence of the primary gas pressure, at constant secondary gas pressure (0.5 MPa), is shown in terms of gas streamlines. Two different operating conditions are to be distinguished, separated by the center line in the figure. On the left of the figure the gas flow field is illustrated with primary and secondary gas flow. One can see a reasonable recirculation flow only close to the primary gas orifice. Entrained gas fom the outside as well as the primary gas stream is flowing with the liquid through the passage of the secondary gas nozzle. No recirculation gas flow underneath the secondary gas nozzle is present. On the right of the figure the gas flow field is illustrated when only the secondary gas (atomization gas) is applied but no primary gas. The shown streamlines illustrate an intensive recirculation flow area located from underneath the secondary gas nozzle up to the primary gas nozzle. Thus, at constant secondary gas pressure the intensity of the gas recirculation underneath the secondary gas nozzle decreases with increasing primary gas pressure. But the intensity of the recirculation gas flow close to the primary gas nozzle increases with increasing primary gas pressure.

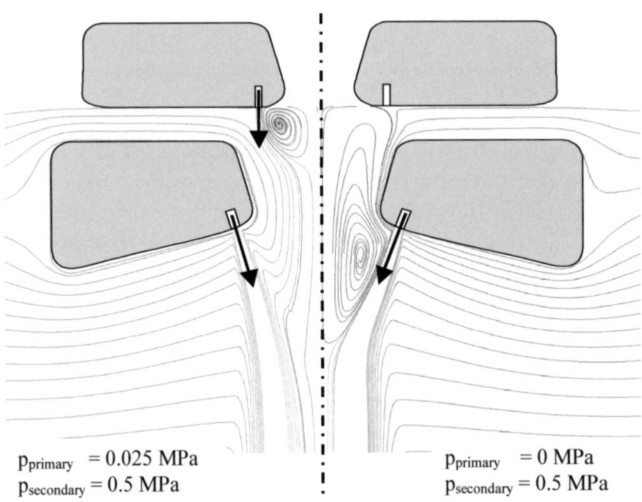

Fig. 3. Flow field of a conventional free-fall atomizer at different operating conditions (From Czisch & Fritsching, 2008a, with permission)

In Figure 4 the axial gas flow velocity on the center line of the atomizer at constant secondary gas pressure (0.5 MPa) is plotted for varying primary gas pressures. For primary gas pressures exceeding 0.025 MPa the recirculation gas flow underneath the secondary gas nozzle is suppressed (no negative velocities in that area). The intensity of the recirculation close to the primary gas nozzle increases with increasing primary gas pressure.

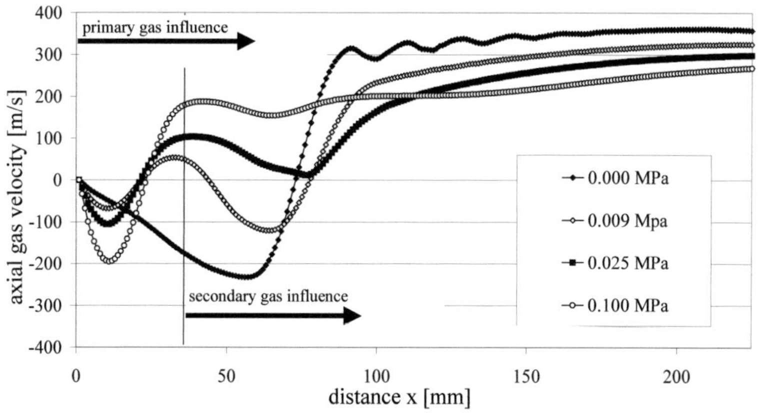

Fig. 4. Axial gas velocity on the center line at varying primary gas pressures (secondary gas pressure 0.5 MPa) (From Czisch & Fritsching, 2008a, with permission)

In Figure 5 the flow field of the free-fall atomizer without primary gas application is illustrated as streamline distribution (left) and local static pressure distribution (right) at a secondary pressure of 0.5 MPa. A recirculation flow field generated underneath the secondary gas nozzle is visible. The area with highest pressure is located where the atomization gas streams impinge onto the center line. The local static pressure exceeds 50 kPa in this area.

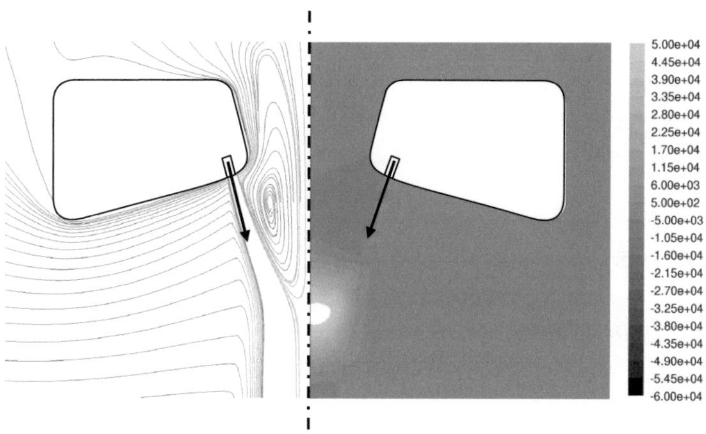

Fig. 5. Flow field (left) and pressure distribution (right) of a conventional free-fall atomizer generated only by the secondary gas flow (scale indicates pressure level in Pa) (From Czisch & Fritsching, 2008a, with permission)

Figure 6 shows the flow situation of a free-fall atomizer without primary gas application but with the installed Coanda-flow ring device at a secondary gas pressure of 0.5 MPa. On the left the flow field is illustrated by its streamline distribution and on the right the static pressure distribution is displayed. No recirculation flow field is generated in this case. On the right it can be seen that at the entrance of the installed Coanda-flow device an underpressure area is generated. Depending on the negative pressure ratio, between the underpressure area close to the atomization zone and the environmental gas pressure, the gas flow is deflected downwards. Thus the mass flow rate increases though the liquid passage of the secondary nozzle suppressing the recirculation completely. The maximum underpressure within the Coanda-flow device exceeds -60 kPa for the given secondary gas pressure.

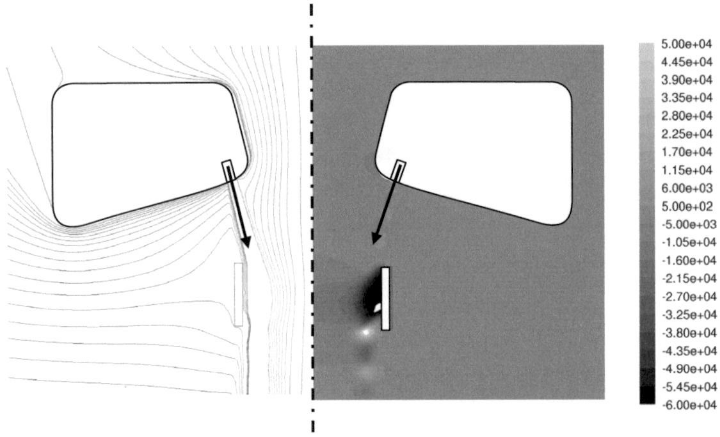

Fig. 6. Flow field (left) and pressure distribution (right) of a free-fall atomizer generated only by the secondary gas flow but with installed Coanda-flow device (scale indicates pressure level in Pa) (From Czisch & Fritsching, 2008a, with permission)

The axial velocity on the center line of the atomizer is plotted in Figure 7 versus the varying secondary gas pressure with usage of the Coanda-flow device. In this investigation no primary gas nozzle is used at all. There is almost no effect on the axial gas velocity with increasing gas pressure until the gas passed the installed flow device. Up to this point the flow is completely dominated by the entrained gas flow from outside. After the gas passed the Coanda-flow device, the axial gas velocity increases with increasing secondary gas pressure. Thus it is possible to achieve a recirculation free flow without applying any primary gas by using the Coanda-flow device.

3.2 Atomization experiments

Model atomization experiments are used to verify the outlined improvements of the operation of the free-fall atomizer with the installed Coanda-flow device. The main aim in this part of the investigation is the proper adjustment of the gas flow rates (or pressures) to ensure stable atomization conditions. Within these model experiments water is atomized by air at room temperature at various pressures (overpressure above the ambient pressure). In Figure 8 (left) a photograph of the atomization area is shown without the Coanda-flow device at 0.5 MPa secondary gas pressure and without primary gas application. Obviously an unstable atomization occurs where droplets are transported from the atomization area upwards towards the atomizer body due to the generated recirculation flow. The instability of the atomization is visible even as three-dimensional structure (asymmetric). When the Coanda-flow device is installed (Figure 8 right) the atomization at 0.5 MPa secondary gas pressure and without primary gas becomes stable, i.e., no major recirculation in the atomization area occurs at all.

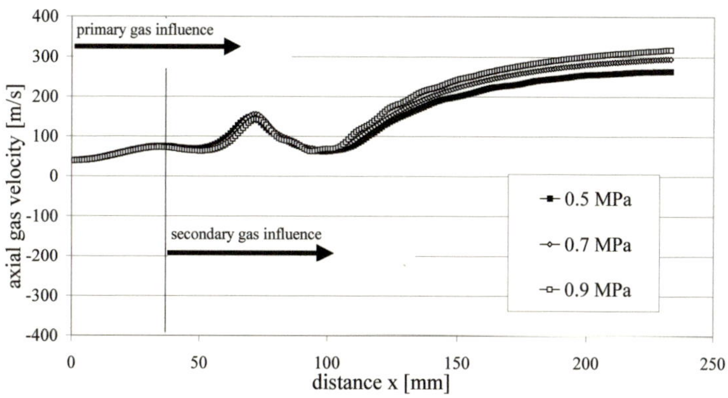

Fig. 7. Axial gas velocity on the center line at varying secondary gas pressures with installed Coanda-flow device (no primary gas nozzle) (From Czisch & Fritsching, 2008a, with permission)

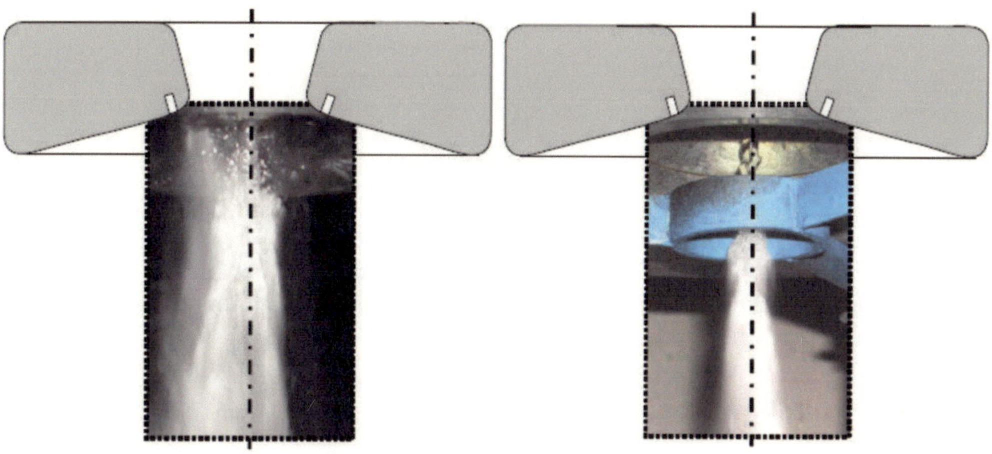

Fig. 8. Model experiments without (left) and with Coanda-flow device (right) (p_{sec} = 0.5 MPa, no primary gas, liquid mass flow rate 435 kg/h) (From Czisch & Fritsching, 2008a, with permission)

The derived Coanda-flow device is adapted within a free-fall atomizer for powder generation from viscous mineral melts. In this application, hot atomization gas is used to suppress fibre formation during viscous melt atomization and to produce spherical particles. Results for achieved particle size spectra have been reported in (Lohner, 2002; Lohner et al., 2005). In Figure 9 the particle size distributions of two different atomization runs are compared at similar operating conditions but with and without the installation of the Coanda-flow device. A mineral melt (liquid melt, no solid particles included) is atomized at a liquid mass flow rate of 300 kg/h. The liquid temperature is app. 1873 K and the gas temperature about 1273 K. The viscosity of the melt at the atomization temperature is 0.2 Pas. The secondary gas pressure used in these cases is similar (0.55 MPa and 0.58 MPa respectively). In the case without the flow device, a primary gas flow is applied at a pressure of 0.18 MPa. This relatively high primary gas pressure allows the suppression of major gas recirculation, but the appearance of the spray structure indicated that some instabilities from the primary gas flow seem to occur. A rather broad particle size distribution results with an indication of a bimodal distribution shape resulting from atomization presumably due to the primary and the secondary gas flows. With the installed Coanda-flow device, a very stable atomizer operation was observed. The resulting mass median droplet diameter (MMD) is app. 280 µm for the conventional free-fall atomizer and app. 140 µm for the free-fall atomizer with the installed flow device. Therefore, the droplet diameter was halved when compared to the conventional free-fall atomizer results.

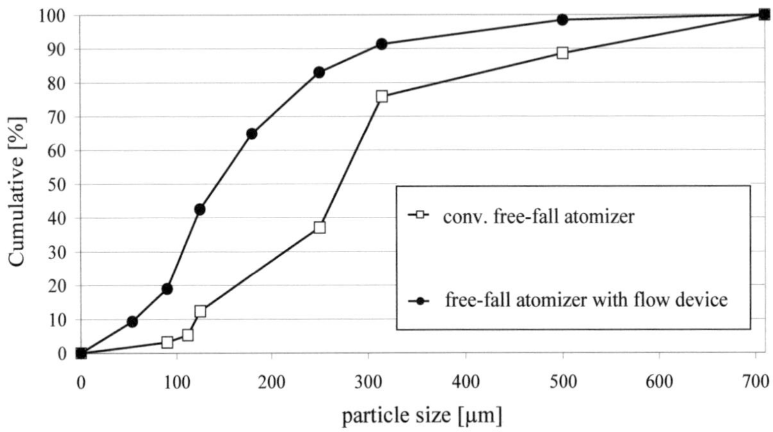

Fig. 9. Particle size distribution with and without applied Coanda-flow device for mineral melt atomization (without Coanda-flow device: p_{sec} = 0.55 MPa, p_{prim} = 0.18 MPa; with Coanda-flow device: p_{sec} = 0.58 MPa, no primary gas; liquid mass flow rate 300 kg/h, liquid viscosity 0.2 Pas, gas temperature 1273 K, melt temperature 1873 K) (From Czisch & Fritsching, 2008a, with permission)

4. Hybrid atomization: Rotary film formation plus gas jet desintegration

Viscous melts, for example glass melts or liquid slags, have a comparably high viscosity and low surface tension. Powder production by gas atomization of such viscous melts is a difficult task because of the rapid cooling of the generated ligaments during melt fragmentation that may cause fibre formation instead of a particulate product. In addition, the increasing temperature of the melt within the process increases the melt viscosity that on the other side decreases the dynamic energy transfer rate between gas and liquid. Therefore, the disintegration time-scale increases and the liquid fragments may be transported out of the area where the most effective atomization occurs. Depending on this decreasing energy-transfer rate, the resulting particle size increases and the efficiency of the atomization process decreases.

By using external mixing gas atomizers with heated atomization gases, in principle a particle product can be obtained (Lohner, 2002; Lohner et al., 2005; Czisch et al., 2003). The efficiency of the atomization process decreases with increasing viscosity and the minimum particle size is limited by the available energy input (Yule & Dunkley, 1994; Strauss, 1999; Dunkley, 2001). Concepts for atomization of highly viscous liquids or melts have already been developed (Pickering et al, 1985; Fraser & Dombrowski, 1962; Campanile & Azzopardi,

2003). Certain limits of these existing technologies are to be seen when realizing an appropriate (low) particle size or a low fibre-to-particle ratio at high melt-mass flow rates.

To increase the efficiency and to decrease the resulting particle size of an atomization process, one can increase the specific surface energy before atomization (Lefebvre, 1980). With respect to twin fluid (gas) atomization, the fragmentation must take place in a region where the velocity difference between the atomization gas and the melt is highest. Thus the melt must be transformed and transported into an area where the maximum velocity difference occurs. To increase the specific surface energy before atomization, a rotating device such as a disc can be used, obtaining high efficiency for transformation. To prevent liquid accumulation effects at the rim of the disc and to decrease the film thickness, the rotary device must operate in sheet-formation mode, generating a free-flowing liquid film from the edge of the disc. The generated film is atomized subsequently by a high-speed gas flow emerging from an external mixing gas atomizer. This configuration allows the generation of a high velocity difference between the atomization gas and the melt film. To prevent fibre formation, the atomization gas needs to be heated. The formation of the free-flowing film is influenced by the local gas flow field. The generated gas flow field can be used to support the transport of the free-flowing film into the most efficient atomization zone. Thus the gas flow field has to be determined by an appropriate atomizer design. An atomizer concept based on the discussed aspects is the prefilming hybrid atomizer.

4.1 The prefilming atomizer concept

The aim of the development is an atomizer design for highly viscous liquids and melts at high throughputs that produce a considerably low droplet size. The prefilming hybrid atomizer introduced here is a combination of a single-fluid rotary atomizer and an external mixing twin-fluid atomizer. The advantage of this specific design is that the feed material is first spread out by a spinning disc due to centrifugal forces, thereby increasing the initial liquid surface prior to the gas atomization process. The rotary atomizer is operated in sheet formation mode so that in the vicinity of the rotary disc a free prefilm is formed. This prefilm is subsequently guided into the atomization zone of the external mixing atomizer by means of intrinsic aerodynamic forces in the gas flow field. In Fig. 10 the principle of the prefilming hybrid atomizer is illustrated. To ensure stable atomization, aerodynamic forces are used to protect the atomizer body from coming into contact and being blocked by parts of the liquid (melt). A sufficiently small gas mass flow rate from the inside of the atomizer protects the atomizer body.

The flow field of gas and liquid of the prefilming hybrid atomizer is illustrated in Fig. 11. Different local flow regimes need to be distinguished. The inner entrainment reaches the atomization area through the atomizer liquid stream passage. The outer entrainment reaches the atomization area directly, and the recirculation gas flow reaches the atomization area from below the rotary disc. All flow regimes are initialized be the atomization gas. The film movement mainly depends on the acting aerodynamic and inertial forces caused by the momentum distribution of the inner entrainment flow, the recirculation gas flow momentum, and the film momentum itself. The aim of the prefilming hybrid atomizer design is to generate a maximum recirculation momentum where the liquid film is transported close to the gas outlet.

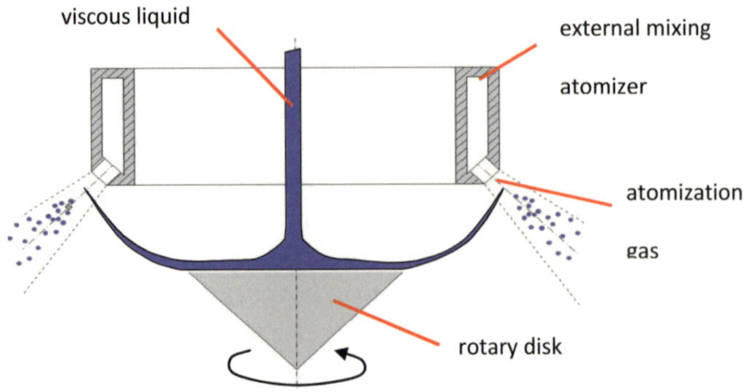

Fig. 10. Sketch of the prefilming hybrid atomizer (From Czisch & Fritsching, 2008b, with permission)

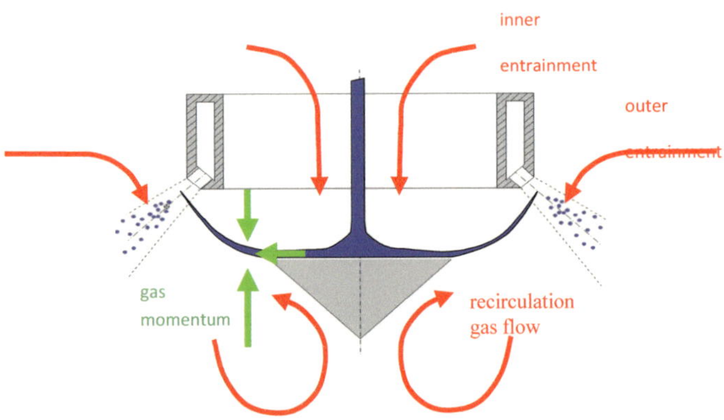

Fig. 11. Gas flowfield and effective momentum transfer of the prefilming hybrid atomizer (From Czisch & Fritsching, 2008b, with permission)

4.2 Numerical simulation

Numerical simulations are used to derive the suitable atomizer design for efficient viscous melt atomization. The aim of the numerical simulations is to derive conditions where a maximum gas momentum on the free flowing liquid film acts against the main axial atomization gas direction. Because the highest velocity difference between gas and liquid and therefore the most efficient atomization area is close to the atomization-gas outlet, the recirculation momentum must be maximized to transport the liquid film towards the outlet of the atomization gas. Within the gas flow simulations the axial momentum of the recirculation gas flow is analyzed at the location of the free-flowing liquid film. In Figure 12 the recirculation momentum depending on the atomization gas outlet angle is shown. The recirculation momentum starts at gas outlet angles above 15°. For outlet angles less than 15° the inner entrainment momentum dominates the interaction with the film. At outlet angles above 15° the recirculation momentum dominates the interaction and it increases with increasing outlet angle. A reasonable high recirculation momentum is reached at outlet angles between 45° to 50°. The atomization gas outlet angle for the design of the hybrid atomizer has therefore been set to 45°.

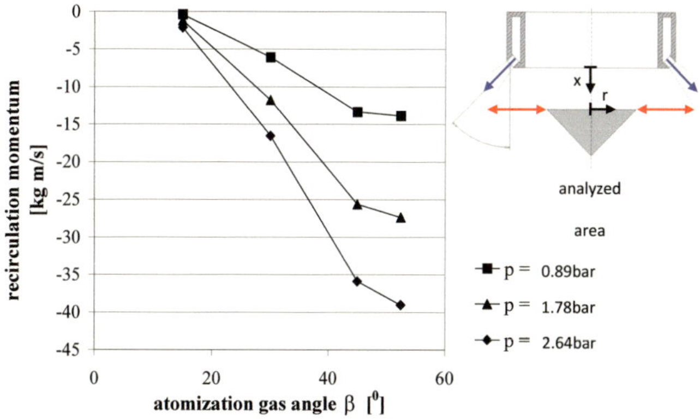

Fig. 12. Axial recirculation momentum on the free flowing film (From Czisch & Fritsching, 2008b, with permission)

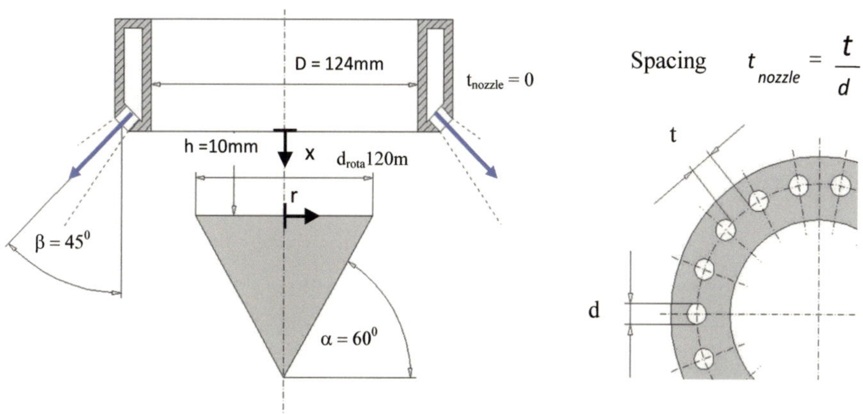

Fig. 13. The constructive parameters of the hybrid atomizer design (From Czisch & Fritsching, 2008b, with permission)

In the same way as the gas outlet angle, all constructive parameters of the atomizer have been determined separately by variation runs within the simulation. In Figure 13 the resulting optimum parameter set for the prefilming hybrid atomizer is shown. The design is based for the atomization of mineral melts at a temperature of 1873 K where the liquid viscosity is 1 Pas and the melt mass flow rate is aimed at 300 kg/h. A gas jet spacing of 0 has been chosen, therefore a slit nozzle is used as gas-flow exit geometry.

4.3 Model experiments

In Figures 14 and 15 experimental results of atomization of water and glycerol are illustrated. The mass median droplet size in the spray measured by laser diffraction is plotted against the air-to-liquid mass-flow ratio (ALR). The efficiency of the hybrid atomizer is evaluated in comparison with a conventional free-fall atomizer (Lohner et al, 2003). The free-fall atomizer is a common device for powder production (Bauckhage & Fritsching, 2000) and spray forming applications (Fritsching and Bauckhage, 2006). When low viscous liquid as e.g. water is atomized, the efficiency of the hybrid atomizer is low. For a constant liquid-mass flow rate, at identical ALR the hybrid atomizer produces coarser particle sizes. In this case the increase of the specific surface energy before atomization is not an advantage. It becomes advantageous only when the viscosity increases. Fig. 15 shows that at constant ALR, the hybrid atomizer produces finer particles than a conventional atomizer for viscous glycerol atomization. In this case, a mass-median particle size of less than 30 μm is achieved.

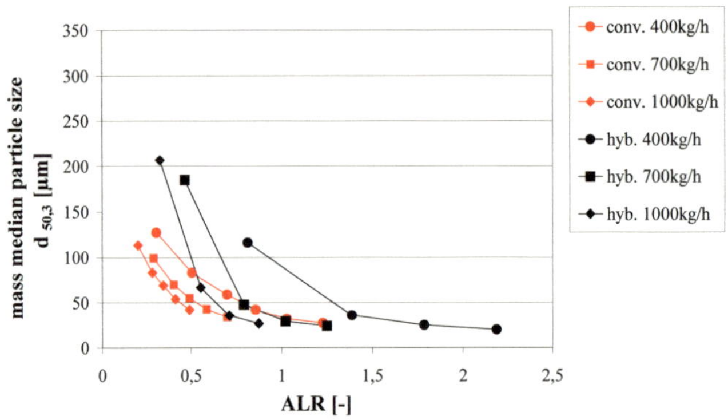

Fig. 14. Experimental results for atomizing water by air (From Czisch & Fritsching, 2008b, with permission)

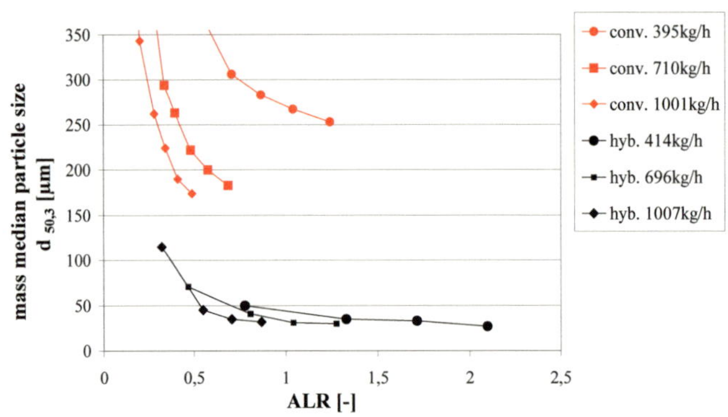

Fig. 15. Experimental results for atomizing glycerol by air (From Czisch & Fritsching, 2008b, with permission)

4.4 Melt atomization experiments

For atomizing viscous melts by the hybrid atomizer concept, a pilot plant for hot-gas atomization has been used (Lohner at al., 2005). The main part of the plant is a spray tower (about 5.5 m in height). On top of the spray tower, the material is melted by an induction heating device. As material to be atomized, a blast-furnace slag is used for the trials. The melt temperature before atomization is measured as 1813-1843 K with a mean temperature of 1826 K. Through the bottom pouring crucible, the melt flows out due to gravity at melt-mass flow rates up to 300 kg/h. The hybrid atomizer located underneath the crucible is operated using heated gas. Gas temperatures up to 1273 K and gas pressures up to 2 bar (relative overpressure against ambient pressure) have been realized within hybrid atomization runs. Hot gas is produced by means of a discontinuous heat exchanger (cowper). The ceramic bulk material inside the cowper is heated by a propane burner. After this heating process, compressed gas or steam is blown through the cowper. The preset values of atomization gas temperature and gas pressure are obtained by mixing the heated gas from the cowper with gas at room temperature. The resulting melt droplets in the spray may be quenched and solidified about 2 m below the atomization nozzle. The solidified powder is collected at the bottom of the spray tower, the fine powder fraction is deposited in the cyclone.

Fig. 16 shows a Scanning-Electron-Micrograph of the sprayed powder from the mineral melt atomization experiment using the hybrid atomization system. A particle fraction of the as sprayed material in the size range from 110 to 350 µm is shown. Mostly spherical particles are to be seen, the fraction of particles to the fraction of fibers in total is above 90 %. In Fig. 16 (right) particles size results for the atomization of this mineral slag melt at different atomization gas temperatures and gas pressures (relative) are shown. The mass-median particle size of the resulting powder is plotted versus the atomization gas temperature. The gas temperature given in the graph is after expansion (a.e.).

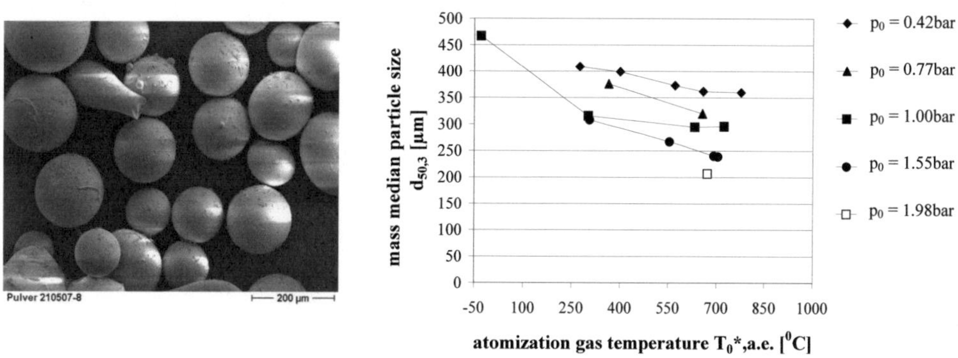

Fig. 16. Micrograph of hot-gas atomized mineral melt particles (size fraction 110 - 350 µm) and results for atomizing mineral melt at an angular disc speed of 1500/min (From Czisch & Fritsching, 2008b, with permission)

The maximum gas temperature before expansion is 1273 K. With increasing gas temperature and increasing atomization gas pressure the particle size decreases. For this set of experimental conditions, the achieved minimum mass median particle size is 210 µm. Compared with the model experiments and the material properties of glycerol the minimum particle size is rather coarse. This results from the limited maximum atomization gas temperature available in the facility. The achieved temperature is still too low in comparison with the melt temperature. Before and within the disintegration process, the melt is cooled down rapidly, so the viscosity increases and the disintegration efficiency is limited.

5. Pressure-gas-atomization

Pressure-gas-atomization is another disintegration process which can be subdivided in two steps: a pre-filming and a gas atomization step. During the pre-filming step, a large molten metal surface is generated to prepare an efficient second disintegration step by gas atomization. In the following, the development, the principle, and the results of the pressure-gas-atomization are described.

5.1 Principle

Pressure-gas-atomization became possible after the development of the pressure atomizer. The idea of a pressure atomizer for molten metal was born in the 90`s by Sheikhaliev and later published by Dunkley and Sheikhaliev (Dunkley & Sheikhaliev, 1995). Originally, the authors called this atomization technique "Centrifugal Hydraulic Atomization". Here, the metal was melted in a crucible placed in a pressure vessel and the molten metal was tangentially fed into a small swirl chamber where the molten metal starts to rotate. Typically, on the centreline of the swirl chamber a gas core develops due to centrifugal forces. Finally, at the lower end of the conical part of the swirl chamber a hole with a diameter between 1 and 2 mm allows the melt to leave the swirl chamber as a thin film creating a hollow cone inside the spray chamber (fig. 17). Normally, a pressure difference of 0.4 to 1.0 MPa between the pressure vessel and the spray chamber is sufficient to achieve a hollow cone of a molten metal. Commonly, this principle is used for cold liquids as an effective atomization system. For a long time, it was doubted if this principle can be applied to molten metal because of their much higher surface tension and the nozzle heating was another challenge. Finally, the molten melt film becomes instable and disintegrates to ligaments and relatively large droplets (mass median diameter far above 100 μm).

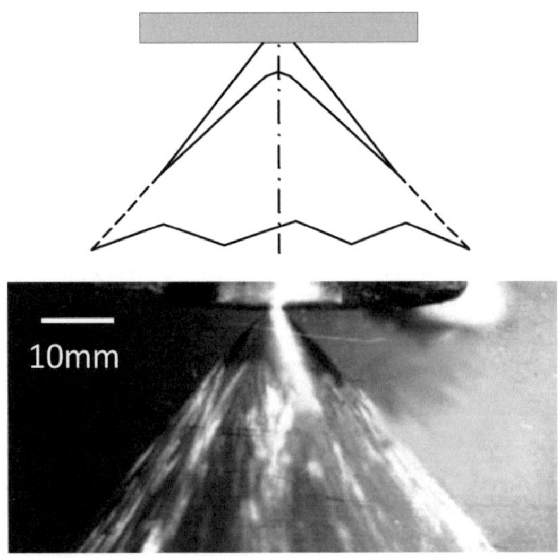

Fig. 17. Schematic (above) and picture (below) of pressure atomization using pure tin (Achelis, 2009).

The pressure-gas-atomization was developed to achieve smaller particles and a narrow size distribution. It combines the pressure atomization as a pre-filming step and the gas atomization. As the second step a gas-ring-atomizer is used to disintegrate the molten metal film, ligaments, and any large droplets. This idea results in a patent published in 2002 (Uhlenwinkel, 2002) and initiated intensive investigations (Lagutkin et al., 2003b, 2004b;

Achelis et al., 2006, 2008, 2009). The hollow cone film/spray is surrounded by the gas-ring-atomizer. Due to the high velocity gas flow from the gas-ring-atomizer, the hollow cone spray from the pre-filming process is atomized into small droplets. Simultaneously, the particles are mainly directed to the centreline of the atomizer axis (Fig. 18).

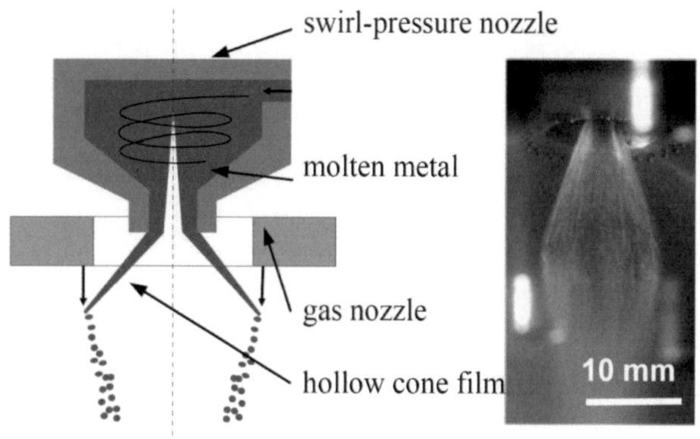

Fig. 18. Schematic (left) and pressure-gas-atomization of pure tin (right), (Achelis et al., 2006a).

5.2 Pre-filming

A pressure difference of about 0.4 MPa is necessary to achieve a fully developed hollow cone if pure tin is used. Fig. 19 shows the effect of the pressure on the cone angle. A fully developed cone is already achieved with a pressure of 0.45 MPa but the cone angle Θ is still small (29°). Higher pressure yields in a large cone angle as demonstrated in the figure as well.

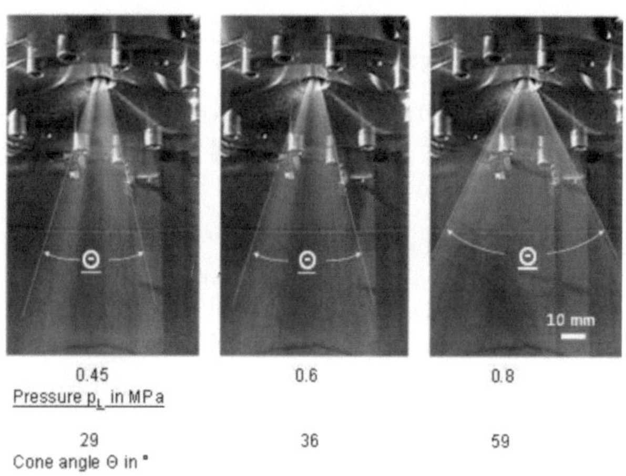

Fig. 19. Effect of the pressure p_L on the spray cone angle Θ during pressure-atomization of pure tin (Achelis, 2009).

A high speed video camera was used to investigate the break-up of the liquid metal film. The pictures in fig. 20 represent single images of different alloys. The image section shows the area below the gas-ring-atomizer (not in use). The tin film is hidden by the gas-ring-atomizer and only the ligaments of the oscillating film are seen below the gas ring nozzle.

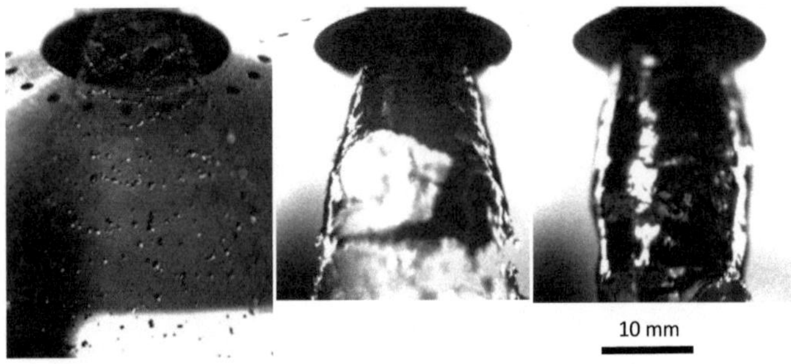

Fig. 20. Break-up of liquid metal films using several alloys. Different film length and break-up mechanism (Achelis et al., 2010).

From the ligaments multiple droplets are generated which are clearly visible below the gas-ring-atomizer. The film break-up of the SnCu30 alloy is totally different. The film can be recognized as the dark area with some bright reflection zones at the surface. The film length is much longer compared to pure tin and the break-up mechanism is dominated by a perforation of the film. The holes in the film grow quickly until only ligaments are left which will break into multiple droplets as well. Finally, the last image (right) shows that the metal film of the AlSi12 alloy is again longer. Bright areas on the surface are just reflections. The spray cone is not fully developed which is indicated by the non-conical shape of the cone. A higher pressure p_L is necessary to achieve a fully developed hollow cone for this alloy.

5.3 Gas flow in the vicinity of the gas-ring-atomizer

The gas flow generated by the gas ring atomizer has been studied carefully for the design of the atomizer. The gas-ring-atomizer can induce strong recirculation zones which can affect the atomization and result in an instable process. Simple pressure measurements on the centreline of the atomizer can help to understand the gas flow in the vicinity of the gas-ring-atomizer. Fig. 21 shows the pressure on a tube surface located on the centreline using different gas-ring-atomizer designs. The location and the intensity of the maximum pressure indicate the risk of recirculation zones. The highest risk is given for the ring slit nozzle with a maximum pressure value of 25 kPa just 10 mm below the gas-ring-atomizer. The other atomizers have several discrete holes with different exit angles and same total exit areas of 18.5 mm^2. An angle of 0° represents the direction straight downwards.

The flow situation was also studied by using water as fluid instead of molten metal. To get an interior view of the spray cone a laser light sheet was used to illuminate the central plane of the spray cone (Fig. 22). White arrows exhibit the exit and direction of the gas jets. Obviously, the ring slit nozzle attacks the water film from the pressure nozzle directly and droplets come very close to the gas nozzle. In case of molten metal this will lead to an instable process. This result corresponds to the pressure measurements in the figure before. Therefore, discrete holes in the ring gas atomizer are preferred for pressure-gas-atomization. An exit angle of 0° was chosen for the atomization of molten metal.

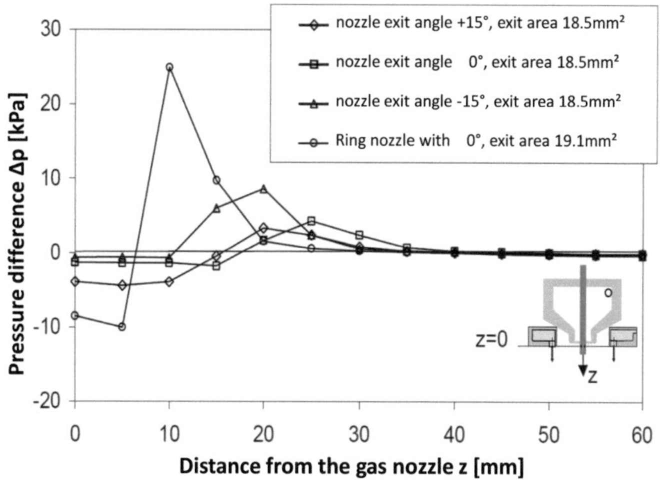

Fig. 21. Pressure on the surface of a tube placed on the centreline of the gas-ring-atomizer indicates the gas flow field in the vicinity of the atomizer. The effect of different atomizer designs is shown (Lagutkin et al., 2003a/b).

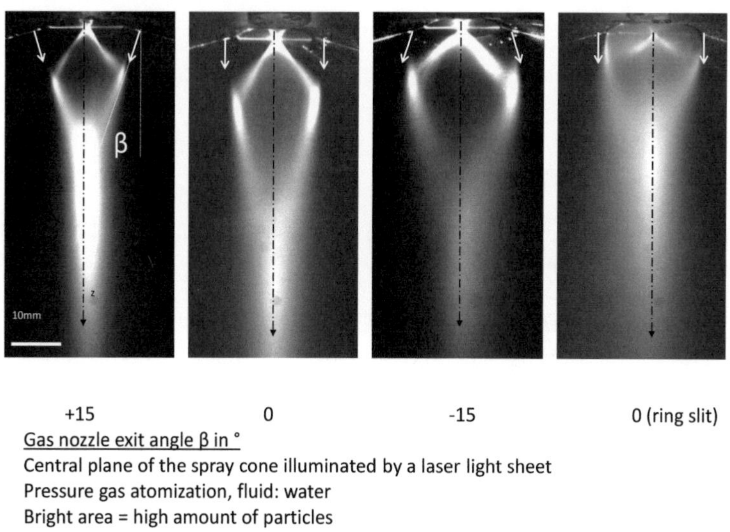

Fig. 22. Influence of different atomizer designs on the spray cone of a pressure-gas-atomizer. The central plane of the spray cone is illuminated by a laser light sheet and water is used as a liquid (Uhlenwinkel et al., 2003).

5.4 Melt flow rate

The production rate is an important feature for an atomization processes. Typically the mass flow of the pressure gas atomization can be varied between 70 and 200 kg/h using tin and tin-copper alloys. As expected, the melt flow rate can be adjusted by the pressure difference between the pressure vessel and the spray chamber. The gas pressure in the gas-ring-atomizer does not influence the melt flow.

5.5 Particle size

Typically, the mass median diameter without gas atomization (only pressure-atomizer) are in the range of 400 to 500 µm. The gas to metal ratio (GMR, here in kg gas / kg melt) is the key parameter to control the mean particle size. Fig. 23 shows that already with low GMR (below 1.0) the mass median diameter can be varied between 40 and 100 µm for pure tin. Even mean particles sizes of approximately 20 µm can be achieved. An increase of the copper content up to 50 weight percent does not make a big difference to the mass median diameter. Only a copper content of 63 weight percent results in larger mean particle sizes. It is assumed the higher surface tension is the main reason for this behavior. All measurements were accomplished by laser diffraction. A superheat of 100 K in the crucible was enough to achieve mostly spherical particles.

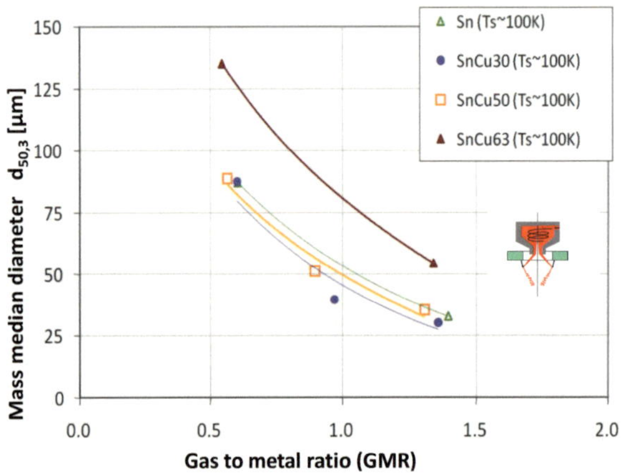

Fig. 23. Effect of the GMR on the mass median diameter for different alloys. The superheat T_S was almost 100K above the liquidus temperature of the alloy (Achelis et al., 2009).

Normally, the size distribution is characterized by another parameter, for example the geometric standard deviation $\sigma_g = d_{84,3}/d_{50,3} = d_{50,3}/d_{16,3}$. Both values are the same if the distribution follows a log-normal size distribution. Here, the standard deviation is defined as $\sigma_g = d_{84,3}/d_{50,3}$. The results in fig. 24 are plotted versus the mass median diameter. Obviously, the geometric standard deviation depends on the mean particle size and a low mean particle size results in a higher geometric standard deviation. This holds for pure tin and this tin-copper alloys as well. A standard deviation below 2.0 achieved by a gas atomization technique is considered extremely good.

Of course, there are other parameters effecting the particle size distribution, specially the design of the atomizers. But the examples shown are representative. The mass median diameter can be calculated by the following semi-empirical equation (Achelis, 2009):

$$d_{50,3} = 1.45\,m \left[0.07 \left(\frac{\sigma_L}{\rho_G u_G^2} \right)^{0.6} \left(\frac{\rho_L}{\rho_G} \right)^{0.1} \delta^{0.4} \left(1 + \frac{\dot{M}_L}{\dot{M}_G} \right) + 0.01 \left(\frac{\eta_L^2 \, \delta}{\sigma_L \rho_L} \right)^{0.5} \left(1 + \frac{\dot{M}_L}{\dot{M}_G} \right) \right] \left(\frac{360}{\Theta} \right)^{0.37} \quad (1)$$

Here, the mass median diameter $d_{50,3}$ (in m) depends on the surface tension σ_L, density ρ_L and viscosity η_L of the melt, the gas density ρ_G, the gas velocity u_G at a distance of 25 mm

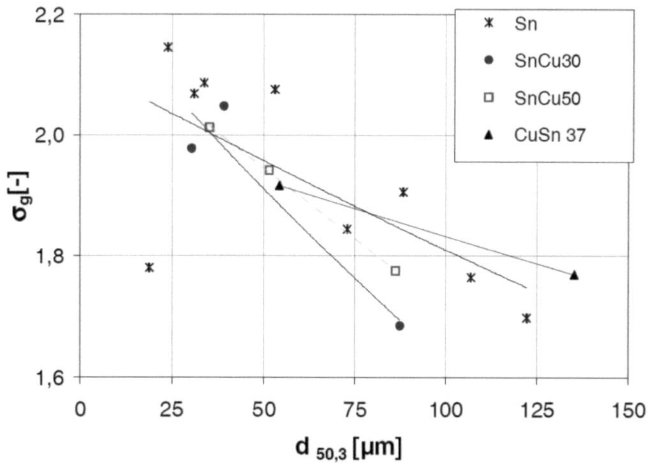

Fig. 24. Geometric standard deviation σ_g versus the mass median diameter $d_{50,3}$ for different alloys to characterize the particle size distribution of a pressure-gas-atomization (Achelis et al., 2010).

from the orifice, the film thickness δ, the melt and gas mass flow $\dot{M}_G$ and $\dot{M}_L$, and the spray cone angle Θ. The film thickness is calculated from the equation given in (Rizk & Lefebvre, 1985 cited in Achelis, 2009) for a pressure atomizer:

$$\delta = 3.66\left(\frac{D_L\,\dot{M}_L\,\eta_L}{\rho_L\,\Delta p_L}\right) \tag{2}$$

In equation (2) D_L means the exit diameter of the nozzle.

5.6 Particle velocity

Particle velocities are of interest for the verification of process modeling and for the calculation of cooling and solidification rates. Several measuring techniques are available to measure the particle velocity in the spray cone (e.g. Laser Doppler Anemometry LDA, Phase Doppler Anemometry PDA, Particle Image Velocimetry PIV). Some results of a PDA measurement are presented in fig. 25. Here, the mean velocity on the centreline of the spray cone is plotted versus the gas pressure in the gas-ring-atomizer at a distance of 165 mm. The molten tin was superheated to a temperature of 275 °C and the melt flow reached 180 kg/h. The variation of the gas pressure resulted in a gas mass flow range between 117 and 198 kg/h. Because of the increased gas flow the mean particle size was reduced and both higher gas velocities and smaller particles added up to mean particle velocity between 70 and 135 m/s.

5.7 Particle shape

Generally, inert gas atomized metal show spherical shape. Therefore, the mean circularity C = 4π projection area / (perimeter)^2 of the particles is close to 1.0. But frequent collisions between liquid or semi-liquid droplets and solidified particles in the spray cone can result in

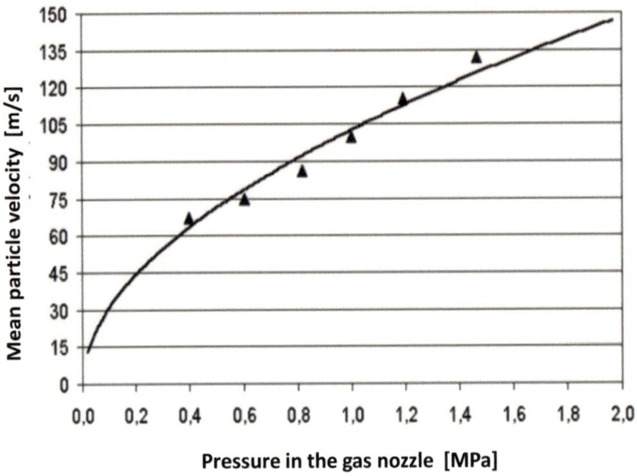

Fig. 25. Correlation between pressure in the gas nozzle and mean particle velocity on the centreline for pure tin measured by Phase-Doppler-Anemometry (Lagutkine et al., 2004a)

agglomerations and/or multiple small particles (satellites) attached to a bigger one. This will reduce the flowability of the powder which is an important quality feature. Often, satellites occur because of the gas flow in the spray chamber. Close to the wall of the spray chamber the gas velocity direction is contrary to the main flow direction and small particles are transported to the spray cone again and collide with the droplets. This recirculation zone inside the spray chamber can be avoided by a recirculation of the clean gas (after the particle collector (cyclone and/or filter) and the fan). The clean gas is added at the top of the spray chamber and - as a side effect - leads to a better visibility of the atomization process.

There appears to be a significant influence of the gas recirculation on the circularity of the powder (figure 26). With gas recirculation the mean circularity of the powder is very close to

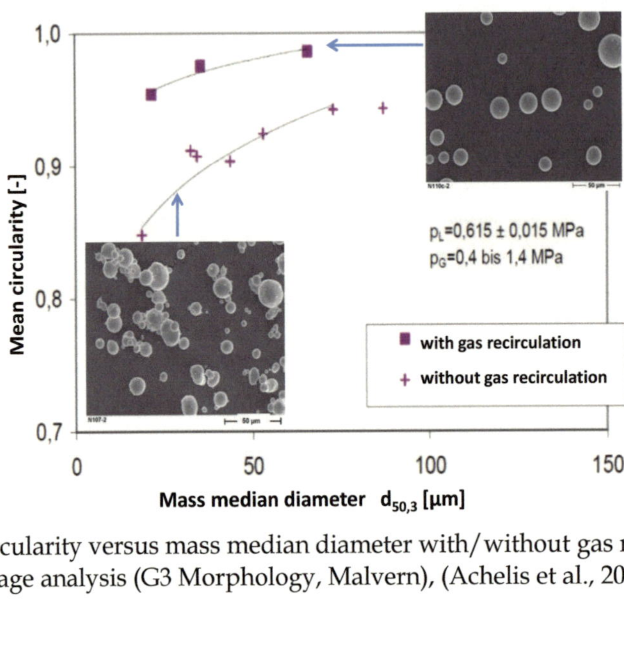

Fig. 26. Mean circularity versus mass median diameter with/without gas recirculation measured by image analysis (G3 Morphology, Malvern), (Achelis et al., 2010)

1.0 and the SEM picture on the right side confirms this result clearly. Without gas recirculation the circularity drops considerably and SEM figure shows the reason of this behavior in agglomerates and multiple satellites.

6. Conclusions and future Investigations

Gas atomization is the most versatile method to produce powders by atomization of a liquid melt. In this context, free-fall atomizers are external mixing gas atomizers based on a combination of a primary and a secondary gas nozzle. The primary gas stabilizes the atomization process and the secondary gas is used for atomization of the liquid. If only the secondary gas is used, a gas recirculation flow underneath the secondary nozzle occurs. This recirculation flow may cause the transport of atomized ligaments or droplets in the direction of the atomizer body. Liquid impinging on the atomizer body may clog the gas orifices and badly influence the atomizer performance. Thus in common free-fall atomizer design, the primary gas is used to suppress the recirculation flow and to stabilize the atomization process.

To improve the operating conditions and to overcome limitations of the free-fall atomizer, a Coanda-flow ring device was developed and installed inside the secondary gas flow. By using the Coanda-flow device, the secondary gas flow is deflected in the downward direction, the entrainment mass flow is increased, and, therefore, a primary gas nozzle is not necessary for suppression of gas recirculation. The concept was derived and verified by means of numerical gas flow simulations and model experiments. The application of the flow-adapted design option for free-fall atomizers was demonstrated by the atomization of a viscous melt. The results indicated a stable operation of the free-fall atomizer even at higher secondary gas pressures, as well as a finer particle yield.

The concept of a prefilming hybrid atomizer has been introduced. The atomizer is designed for atomizing highly viscous liquids and melts. Physical gas flow effects are utilized to increase the atomizer efficiency. The hybrid atomizer design is based on numerical flow simulations of the gas phase. Experiments where water a viscous model liquid (glycerol) a have been atomized with air show a higher disintegration efficiency than a conventional external mixing atomizer for the viscous liquid atomization process. Model experiments atomizing fluids of low viscosity like water show an insufficient atomization efficiency of the hybrid atomizer design. Thus the hybrid atomizer concept is a novel device for highly viscous liquids and melts. The experimental results for atomization of a mineral melt by heated gases underline the success of the hybrid atomization process with some limitations. For the present conditions, a minimum mass-median particle size of 210 μm has been achieved. The coarse product obtained by melt atomization is due to limitations of the available maximum atomization gas temperature and may be overcome in future developments.

The complexity of the pressure-gas-atomization is a drawback for frequent applications in industry. To overcome the problem the system can be simplified in the future if it is used with low melting point alloys. Today, pumps for molten metal like tin alloys are available and the complexity of the system can be reduced considerably because a pressure vessel is not necessary anymore and thus the process principally can operate continuously.

Another future objective is the use of the pressure-gas-atomizer for coating applications. It has already been proofed that this atomization system is dedicated to generate thick coatings on a tube. As an example a steel tube with a diameter of the 80 mm was coated with a 10 mm thick tin layer. The density of the coating was much better the than 99% of the full density (Uhlenwinkel et al., 2008).

7. Acknowledgments

The results that have been presented here are mainly based on the project work of several PhD students at the Particles and Process Engineering department of the University of Bremen. Namely the contributions of Dr.-Ing. L. Achelis, Dr.-Ing. S. Markus, Dr.-Ing. H. Lohner, Dr.-Ing. U. Heck, Dr. S. Pulbere and Dr.-Ing. C. Czisch are acknowledged. The cooperation in these fields with international colleagues, namely Dr. S. Lagutkin, Prof. S. Sheikhaliev and Dr. V. Srivastava is greatfully acknowledged.

The financial support of the projects has been mainly given by the German Research Foundation (DFG), whose support during the past decade in several areas and through projects is greatfully acknowledged.

8. References

Achelis, L.; Uhlenwinkel, V.; Lagutkin S. & Sheikhaliev, S. (2006): Atomization Using a Pressure-Gas-Atomizer. *Proceedings of the International Conference on Powder Metallurgy 2006*, Busan (South Koria), Sept. 2006

Achelis, L. & Uhlenwinkel, V. (2008): Characterisation of metal powders generated by a pressure-gas-atomizer. *MSE A* 477 (2008), pp.15-20

Achelis L. (2009): *Kombinierte Drall-Druck-Gaszerstäubung von Metallschmelzen*. Shaker Verlag, Aachen 2009, ISBN 978-3-8322-8012-3

Achelis, L.; Sulatycki, K.; Uhlenwinkel, V. & Mädler, L. (2010): Spray angle and particle size in the pressure gas atomization of tin and tin-copper alloys, *Proceeding of the International Conference on Powder Metallurgy 2010*, Florence (Italy), October 2010

Anderson, I.E. & Figliola, R.S. (1998). Observations of gas atomization process Dynamics, in *Modern Developments in Powder Metallurgy*, Gummeson, P.U. &Gustafson, D.A. Eds., Metal Powder Industries Federation, Princeton, N.J., Vol. 20, pp. 205–223

Ashgriz, N. (2011). *Handbook of Atomization and Sprays*, Springer, New York

Bayvel, L. & Orzechowski, Z. (1993). *Liquid Atomization*, Taylor & Francis ,Washington

Bauckhage, K. (1992). Das Zerstäuben metallischer Schmelzen, *Chem.-Ing.-Tech.*, Vol. 64 No. 4, pp. 322–332

Bauckhage, K. & Fritsching U. (2000). in: K.P. Cooper, I.E. Anderson, S.D. Ridder, F.S. Biancaniello (Eds.) *Liquid Metal Atomization: Fundamentals and Practice*, TMS, Warrendale, USA, pp. 23 - 36

Bergmann, D.; Fritsching, U. & Bauckhage, K. (2001) Simulation of molten metal droplet sprays, *Comp. Fluid Dynamics-Journal*, Vol. 9, pp. 203 - 211

Bradley, D. (1973). On the atomization of liquids by high-velocity gases, Part 1, *J. Phys. D: Appl. Phys.*, Vol. 6, 1724–1736, Part 2, *J. Phys. D: Appl. Phys.*, Vol. 6, pp. 2267–2272

Campanile, F. & Azzopardi, B.J. (2003). in: Cavaliere, A. (Ed) CD-ROM *Proc. International Conference on Liquid Atomization and Spray Systems ICLASS 2003*, Sorrento, Italy, 13-17.07.2003, ILASS-Europe

Czisch, C.; Lohner, H.; Fritsching, U., Bauckhage, K. & Edlinger, A. (2003). in: K. Bauckhage, U. Fritsching, J. Ziesenis, A. Uhlenwinkel, A. Leatham (Eds), *Proc. Spray Deposition and Melt Atomization Conf. SDMA 2003*, Bremen, 22.-25.6.2003

Czisch, C.; Lohner, H. & Fritsching, U. (2004). Einfluss der Gasdüsenanordnung auf den Desintegationsvorgang und das Zerstäubungsergebnis bei der Zweistoff-zerstäubung, *Chem.-Ing.-Tech.*, Vol. 76, No. 6, pp. 754-757

Czisch, C. & Fritsching, U. (2008a). Flow-adapted design for Free-fall atomizers, *Atomization and Sprays*, Vol. 18 No. 6, pp. 511-522

Czisch, C. & Fritsching, U. (2008b). Atomizer design for viscous-melt atomization, Mat. Sci. Engng. A Vol. 477 No. 1-2, pp. 21-25

Dombrowski, N. & Johns, W.R. (1963). *Chem. Eng. Sci.*, Vol. 3, pp. 203-214

Dunkley, J.J. & Sheikhaliev, S. (1995): Single Fluid Atomization of Liquid Metals, *Proceedings of the International Conference on Powder Metallurgy & Particulate Materials*, Seattle (USA), May 1995, Vol. 1, pp. 79-87

Dunkley J.J. (2001). in: 2001 International Conference on Powder Metallurgy and Particulate Materials PM²TEC 01, 2-29-2-35, 2001, Metal Powder Industries Federation, Princeton, USA

Fraser, R.P.; Dombrowski, N. & Routley, J.H. (1962). *Chem. Eng. Sci.*, Vol. 18, pp. 339-353

Fritsching, U. & Bauckhage, K. (1992). Investigations on the atomization of molten metals: The coaxial jet and the gas flow in the nozzle near field, *PHOENICS J. Comp. Fluid Dynamics*, Vol. 5, No. 1, pp. 81-98

Fritsching, U. (2004). *Spray Simulation: Modeling and Numerical Simulation of Sprayforming Metals*, Cambridge University Press, Cambridge, UK

Fritsching, U. & Bauckhage, K. (2006a). *Sprayforming of Metals*, in: Ullmann´s Encyclopedia of Industrial Chemistry, 7th Edition, 2006 Electronic Release, Wiley VCH, Weinheim, Germany

Fritsching, U. (2006b). Spray Systems, in: Multiphase Flow Handbook, Ed.: C.T. Crowe, CRC-Press, Boca Raton, Fl, USA, 2006, Chapter 8, pp. 8.1 - 8.100; ISBN: 0849312809

Heck, U. (1998). *Zur Zerstäubung in Freifalldüsen*, VDI Verlag GmbH Düsseldorf, Germany

Heck, U.; Fritsching, U. & Bauckhage, K. (2000). Gas-Flow Effects on Twin-Fluid Atomization of Liquid Metals, *Atomization and Sprays,* Vol. 10, No. 1, pp. 25 – 46

Lagutkin, S.; Achelis, L.; Sheikaliev, S.; Uhlenwinkel, V. & Srivastava V. (2003a): Atomisation Process for Metal Powder. *Proceedings of the International Conference on Spray Deposition and Melt Atomization 2003*, Bremen (Germany), June, 2003,

Lagutkin, S. (2003b): *Development of Technology and Equipment for Metal Powder Production by Centrifugal-Gas Atomization of Melt.* Ph.D. thesis. Ekaterinburg, Ural Department of Academy of Sciences. 2003, in Russian

Lagutkin, S.; Uhlenwinkel, V.; Achelis, L.; Pulbere, S. &. Sheikhaliev S. (2004a): Centrifugal-Gas Atomization: Preliminary Investigation of the Method, *Proc. International Conference on Powder Metallurgy 2004*, Wien, Austria, Oct. 2004, Vol. 1, S.71-76

Lagutkin, S.; Achelis, L.; Sheikhaliev, S.; Uhlenwinkel, V. & Srivastava V. (2004b): Atomization process for metal powder. *MSE-A* 383 (1), (2004), pp. 1-6

Lavernia, E.J. & Wu, Y. (1996). *Spray Atomization and Deposition*, J. Wiley & Sons, Chichester

Lefebvre, A.H. (1980). *Prog. Energy Combust. Sci.*, Vol. 6, pp. 233-261

Lefebvre, A.H. (1989). *Atomization and Sprays*, Hemisphere Publ. Corp., US

Liu, H. (2000) *Science and Engineering of Droplets: Fundamentals and Applications,* William Andrew Publ., Norwich, USA

Lohner, H. (2002). *Zerstäuben von Mineralschmelzen mit Heißgas,* PhD. thesis, University Bremen

Lohner, H.; Czisch, C. & Fritsching, U. (2003). Impact of Gas Nozzle Arrangement on the Flow Field of a Twin Fluid Atomizer with External Mixing, *Int. Conf. on Liquid Atomization and Spray Systems*, Sorrento, Sep. 2003 (Italien)

Lohner, H.; Czisch, C.; Schreckenberg, P.; Fritsching, U. & Bauckhage, K. (2005). Atomization of viscous melts, *Atomization and Sprays*, Vol. 15, No. 2, pp. 169-180

Markus, S.; Fritsching, U. & Bauckhage, K. (2002). Jet Break Up of Liquid Metals in Twin Fluid Atomization, *Materials Sci. & Engng. A*, No. 326, pp. 122 - 133

Pickering, S.J.; Hay, N.; Roylance, T.F. & Thomas, G.H. (1985). *Ironmaking and Steelmaking,* Vol. 12 No. 1

Rizk, N.K. & Lefebvre A.H. (1985) : Internal Flow Characteristics of Simplex Swirl Atomizer. *Journal of propulsion and power*, Vol. 1 (1985) No. 3, New York, pp. 193-199

Strauss, J.T. (1999). Hotter gas increases atomization efficiency, *Metal Powder Report,* Vol. 11, pp. 24-28

Uhlenwinkel, V. & Sheikhaliev S., et al. (2002): *Verfahren und Vorrichtung zum Herstellen von Metallpulver und keramischem Pulver*, Pat.Nr. 10237213, 14.Aug. 2002

Uhlenwinkel, V.; Achelis, L.; Sheichaliev, S. & Lagutkin, S. (2003): A new Technique for Molten Metal Atomization. *Proc. ICLASS 2003*, Sorento, Italy, June, 2003

Uhlenwinkel, V.; Achelis, L. & Mädler, L. (2008): New atomization process to achieve high production rates during thermal spraying of thick coatings, Proceedings of the *International Thermal Spray Conference 2008*, Maastricht (Netherlands), June 2008

Wille, R. Fernholz, H. (1965). Report on the First European Mechanics Colloquium on the Coanda Effect, *J. Fluid Mech.*, Vol. 23, pp. 801-819

Yule, A.J. & Dunkley J.J. (1994). *Atomization of Melts*, Clarendon Press Oxford, UK